Cloud Computing

Cloud Computing
Concepts and Technologies

By
Dr. Sunilkumar Manvi
Dr. Gopal K. Shyam

CRC Press
Taylor & Francis Group
Boca Raton London New York

CRC Press is an imprint of the
Taylor & Francis Group, an **Informa** business

First edition published [2021]
by CRC Press
6000 Broken Sound Parkway NW, Suite 300, Boca Raton, FL 33487-2742

and by CRC Press
2 Park Square, Milton Park, Abingdon, Oxon, OX14 4RN

ISBN: 9780367554590 (hbk)
ISBN: 9780367554613 (pbk)
ISBN: 9781003093671 (ebk)

Typeset in Computer Modern font
by KnowledgeWorks Global Ltd

To All Our
Beloved Ones.

Contents

Foreword

As Cloud continues to revolutionize applications in academia, industry, government, and numerous different fields, there are some genuine difficulties at both theoretical and practical levels that will regularly require new approaches and practices in all domains. Comprehensive and timely, this book titled, *Cloud Computing: Concepts and Technologies*, details progress in Cloud computing and offers bit by bit guidance on the best way to understand and implement it.

Motivation for This Book

Gartner, in a February 2, 2009 press release, posted the question of why, and when "the Cloud computing market is in a period of growth and high potential but still requires several years and many changes in the market before it becomes a mainstream IT effort." However, this concern, while valid, is not insurmountable. We already have many examples of successful Cloud computing implementations, from small organizations up to large enterprises, such as the U.S. Department of the Navy. The National Institute of Standards and Technologies (NIST) released its first guidelines for agencies that want to use Cloud computing, and groups such as Jericho forum brought security executives together to collaborate and deliver solutions.

The issues, however, does exist with regard to data ownership rights, performance and availability. While these are all valid concerns, solutions do exist and are being fine-tuned everyday; the challenge is in bringing executives out of a state of unawareness and fear, giving them the understanding and knowledge necessary to make informed, educated decisions regarding their Cloud initiatives.

This book is an attempt by us to educate the readers and novice researchers with a proper understanding of Cloud computing. The book offers a thorough and detailed description of Cloud computing concepts, architectures, and technologies. It serves as a great reference for both newcomers and experts and is a must-read for any IT professional interested in Cloud computing.

About the Authors

Dr. Sunilkumar S. Manvi is currently working as a Professor and Director in School of Computing and Information Technology, REVA University, Bengaluru, India. He received his B.E. Degree from Karnataka University, M.E. degree in Electronics from the University of Visweshwariah College of Engineering, Bangalore and Ph.D. degree in Electrical Communication Engineering, Indian Institute of Science, Bangalore, India. He has vast experience of more than 30 years in teaching and research. His research interests are: Software Agent based Network Management, Wireless Networks, Multimedia Networking, Underwater Networks, Wireless Network Security, Grid and Cloud computing. He has published around 295 papers in international journals and conferences and 15 publications in books/book-chapters. He is a Fellow IETE India, Fellow IEEE India, Member ACM USA and senior member of IEEE USA. He received best research publication award from VGST Karnataka in 2014. He has been awarded with Sathish Dhawan Young Engineers State Award for outstanding contribution in the field of Engineering Sciences in 2015 by DST Karnataka India. He is a reviewer and programme committee member for many journals.

Dr. Gopal K. Shyam is currently working as a Professor in School of Computing and Information Technology, REVA University, Bengaluru, India. He received B.E. M.Tech, and Ph.D from Visvesvaraya Technological University, Belagavi. He has handled several several subjects for UG/PG Students like Algorithms, Computer Networks, Web programming, Advanced Computer architecture, Information security, Computer Concepts and C Programming. His research interest includes Cloud computing, Grid computing, High performance computing, etc. He is a member of IEEE and ACM and is actively involved in motivating students/faculties to join professional societies. He has an experience of around 13 years in Teaching and Research.

Preface

With the advancements in Information Technology, we are witnessing tremendous development in computer and communication technologies. There has been a lot of work happening in the area of Cloud computing. The Cloud users are demanding various resources, viz. software, platform and infrastructure with better quality of services from the Cloud service providers. Hence, it has become essential for engineers, scientists and professionals to know about Cloud computing and underlying technologies.

The book provides an explanation on the Cloud computing concepts, architectures, functionalities, technologies and applications. It covers the multicore architectures, distributed and parallel computing models, virtualization, Cloud developments, workload and Service-Level-Agreements (SLA) in Cloud, resource management, issues etc. This book is ideal for a broad audience that includes researchers, engineers, IT professionals and graduate students. This book comprises of twelve chapters followed by laboratory setups and experiments. Each chapter has Multiple Choice Questions (MCQs) questions with answers, review questions and critical thinking questions. The contents of each chapter has a smooth flow of contents provided through practically-focused topics.

Organization of the Book

The topics discusssed in various chapters are as follows:

- Chapter 1 gives a brief history of Cloud computing. Basic terminology and concepts are introduced, along with descriptions of common benefits and challenges of Cloud computing adoption and a discussion of business drivers and technology innovations. Cloud delivery and Cloud deployment models are also elaborated.

- Chapter 2 focuses on discussion of parallel and distributed systems, and how these technologies differ from the conventional centralized computing systems.

- Chapter 3 deals with multicore architectures for Cloud computing environment, and exploiting the parallelism in the hardware and software.

- Chapter 4 discusses the heart of Cloud computing i.e virtualization, virtualization techniques, virtualization technologies e.g., Xen, VmWare etc. The Pros and Cons of virtualization is also covered.

- Chapter 5 covers description of Infrastructure as a Service (IaaS) in Cloud. Here we discuss virtual machines provisioning and migration services for management of virtual machines which helps in capacity planning in Cloud, conserving energy efficiency in Clouds.

- Chapter 6 considers PaaS and SaaS business model in Cloud.

- Chapter 7 deals with capacity planning in Cloud.

- Chapter 8 discusses SLA management in Cloud computing.

- Chapter 9 discusses resource management in Cloud.

- Chapter 10 covers Cloud computing development tools.

- Chapter 11 focuses on Cloud security.

- Chapter 12 discusses research issues for novice researchers.

Appendix chapters 13 and 14 cover experiments on CloudSim and Cloud platforms, respectively.

Why the Book Is Different

- This book lays a good foundation to the core concepts and principles of Cloud computing, walking the reader through the fundamental ideas with expert ease.

- The book advances on the topics in a step-by-step manner and reinforces theory with a full-fledged pedagogy designed to enhance students' understanding and offer them a practical insight into the subject.

- There are chapters on resource allocation issues and security, which is widely talked in research domain.

- Includes seperate chapters on technical and legal issues in Cloud.

- Provides CloudSim-based simulations and real Cloud platforms based experiments for practical experience.

Acknowledgments

We are very much thankful to the almighty for giving us motivation, confidence and blessings for the completion of the book.

We owe our deepest gratitude and sincere thanks to Dr. P. Shyamaraju, Chancellor, REVA University, for creating wonderful infrastructure and incredible academic ambiance that facilitated us to bring out this book in its current form. We would like to express our special thanks to honorable Vice-Chancellor Dr. K. Mallikharjuna Babu, Provost Dr. S. R. Shankpal, Registrar Dr. M. Dhanamjaya, for providing us the infrastructure to carry out book writing project.

Our special thanks to Dr. P. I. Basarkod and Dr. P. Nagesha who have given their constant support and valuable suggestions throughout my course of book writing. Our thanks to office and library staff of REVA University, Bangalore, for the kind support during our work.

We thank our contemporary researchers of Wireless Information Systems Research Laboratory, Dr. Shrikant S. Tangade and Mr. Nirmal Kumar Benni for their constant support and encouragement that yielded useful discussions and evolved into fruitful research outcomes. It is a great experience to work with such a talented research team.

We are also greatly thankful to our family members, Bharathi Manvi, Sanket Manvi, Priyanka Bharti, Prayuta Krishna, relatives and friends for their love and moral supportfor successful completion of the book. We thank our parents for their encouragement.

Finally, we thank our publisher, CRC, for their continuous support. We welcome all comments and suggestions for future enhancement of the book.

Authors

Dr. Sunilkumar S. Manvi
Professor & Director,
School of Computing & Information
Technology,
REVA University, Bengaluru, India

Dr. Gopal K. Shyam
Professor,
School of Computing & Information
Technology,
REVA University, Bengaluru, India

Symbols

Symbol Description

$Sp(n)$ Speedup

$E(n)$ Efficiency

C_i Capacity of ith autonomic service

$N(t)$ Actual number of autonomic services allocated at a given time

$M(t)$ Number of autonomic services to be added/removed at a given time

C^{av} Average capacity among all system autonomic services

$L_i(t)$ Load of autonomic service i computed at time t over timeframe $(t - T^m, t)$ (percent with respect to autonomic service capacity)

$L^{av}(t)$ Average load per autonomic service of the system, computed at time t over timeframe $(t - T^m, t)$ for all autonomic services in the system (percent)

$\tilde{L}_i^{av}(t)$ Average load per autonomic service of the neighborhood of autonomic service i, computed at time t over timeframe $(t - T^m, t)$ (percent)

L^{min} Minimum load threshold (percent)

L^{max} Maximum load threshold (percent)

L^{des} Desired load threshold, which is equal to $(L^{max} - L^{min})/2$ (percent)

T^s Period between two successive runnings of the auto-scaling algorithm on a autonomic service

T^m Length of monitoring timeframe for the actual load

$neig_i$ Neighborhood of autonomic service i: contains autonomic service i and its neighbors

$queue_i$ Number of enqueued requests in autonomic service i

R_{max} Maximum response time for completed requests (from SLA)

1

Introduction

Learning Objectives
After reading this chapter, you will be able to

- Define the limitations of traditional technologies

- Appraise the concept of Cloud computing

- Discuss the applications, tools and technologies in Cloud computing

Three decades prior, the personal computers (PCs) brought change in the manner in which individuals worked. Yet, over the ongoing years another technology called the Cloud has pulled in research network since it guarantees to address different information technology (IT) challenges. Physical areas for asset access, gigantic expense to benefit interminable power, improvement in the courses of events for applications have bothered IT and its customers since long time. Cloud computing guarantees to offer access to gigantic measures of information and computational power, empowering creators to take care of issues in a better approach to accomplish a dependable IT foundation.

The Cloud framework can be seen as containing both a physical layer and a virtualization layer. The physical layer comprises of the equipment assets that are important to help the Cloud administrations, and regularly incorporates server, stockpiling and system parts. The virtualization layer comprises of the product sent over the physical layer which shows the fundamental Cloud attributes. Reasonably, the virtualization layer is positioned over the physical layer.

This chapter presents Cloud service delivery models, characteristics, benefits, platforms and technologies.

1.1 Cloud Computing

A Cloud is defined as a space over network infrastructure where computing resources such as computer hardware, storage, databases, networks, operating systems, and even entire software applications are available instantly, on-demand. It is true that Cloud computing may not involve a whole lot of new technologies, but the fact that it surely represents a new way of managing IT cannot be denied. For example, scalability and cost savings can be achieved to the largest extent from Cloud computing.

Cloud computing is bound to be compared with service oriented architectures (SOA), Grid computing , Utility computing and Cluster computing. Cloud computing and SOA are persued independently. Platform and storage services of Cloud computing offers value addition to SOA's efforts. With technologies like Grid computing, computing resources can be provisioned as a utility. Whereas, Cloud computing goes a step further with on-demand resource provisioning. It also removes the necessity of over-provisioning to accomodate the demands of several customers. Utility computing is paying for resource usage, similar to the way we pay for a public utility (such as electricity, gas, and so on).

Cluster computing is a cheaper-cost form for processing applications that can run in parallel. The summary of the features of each of these computing techniques is listed in Table 1.1. Figure 1.1 depicts the construction and deployment for Cloud service delivery model. End-users access Cloud services such as computing and datacenter resources via the Internet. The user needs to have an account with the Cloud service provider (CSP) for security and billing schemes. The required resources are specified by the users. The CSP provisions resources in the form of virtual machines directly to user accounts. This offering facilitates users more flexibility in building their own applications on top of remotely hosted resources. Users essentially rent operating systems, CPU, memory, storage and network resources from the CSP to improve elasticity and scalability of workloads.

TABLE 1.1

Computing techniques

Computing techniques	Features
Cloud computing	Cost efficient, almost unlimited storage, backup and recovery, easy deployment
SOA	Loose coupling, distributed processing, asset creation
Grid computing	Efficient use of idle resources, modular, parallelism can be achieved, handles complexity
Cluster computing	Improved network technology, processing power, reduced cost, availability, scalability.

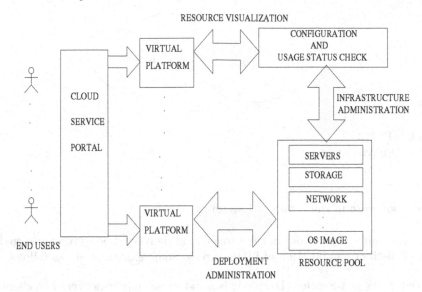

FIGURE 1.1
Cloud scenario

1.2 Service Delivery Models

A Cloud delivery model represents a specific, pre-packaged combination of IT resources offered by a Cloud provider. Three common Cloud delivery models (IaaS, PaaS, SaaS) which have become widely established and formalized are explained as follows:

Infrastructure as a Service (IaaS): The capability provided to the consumer is to provision processing, storage, networks, and other fundamental computing resources. Example: running CPU/memory intensive application using Amazon IaaS Cloud.

Platform as a Service (PaaS): The capability provided to the consumer is to deploy onto the Cloud infrastructure consumer-created or acquired applications using programming languages, libraries, services and tools supported by the provider. Example: building and deployment of application using Google Cloud Platform.

Software as a Service (SaaS): The capability provided to the consumer is their applications running on a Cloud infrastructure. Example: opening word/PDF files using Google Apps without installation of MS Office/Adobe

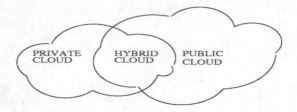

FIGURE 1.2
Cloud deployment models

Reader software in the networked system.

Many specialized variations of the three Cloud delivery models have emerged, each of distinct combination of IT resources. Some examples are as follows.

- Database-as-a-Service (DaaS): It is a database that is delivered to clients via a Cloud computing platform. Access to it is provided as a service. Database services take care of scalability and high availability of the database.

- Communication-as-a-Service (CaaS): It is an outsourced enterprise communications solution that can be leased from a single vendor. Such communications can include voice over IP (VoIP or Internet telephony), instant messaging (IM), collaboration and videoconference applications using fixed and mobile devices.

- Integration-Platform-as-a-Service (IPaaS): It is a Cloud service that provides a platform to support application, data and process integration projects, usually involving a combination of Cloud-based applications and data sources, APIs and on-premises systems.

- Testing-as-a-Service (TaaS): It is related to the outsourcing of testing activities to a third party that focuses on simulating real-world testing environments as specified in the client requirements. In other words, TaaS is an outsourcing model.

1.3   Deployment Models

Cloud hosting deployment models represent the exact category of Cloud environment and are mainly distinguished by the proprietorship, size and access. It tells about the purpose and the nature of the Cloud. In order to know which deployment model matches any of the workload requirements, it is necessary to know the four deployment models (shown in Figure 1.2), discussed as follows.

Public Cloud is a type of Cloud hosting in which the Cloud services are delivered over a network which is open for public usage. This model is a true representation of Cloud hosting ; in this, the service provider renders services and infrastructure to various clients. Hybrid Cloud is a type of Cloud computing, which is integrated. It can be an arrangement of two or more Cloud servers, i.e. private, public or community Cloud that is bound together but remain individual entities.

Private Cloud is the platform that is implemented on a Cloud-based secure environment that is safeguarded by a firewall which is under the governance of the IT department that belongs to the particular corporate. Private Cloud permits only the authorized users, gives the organisation greater and direct control over their data.

Community Cloud is a type of Cloud hosting in which the setup is mutually shared between many organisations that belong to a particular community, i.e. banks and trading firms. It is a multi-tenant setup that is shared among several organisations that belong to a specific group which has similar computing apprehensions.

1.4 Characteristics and Benefits of Cloud Computing

Cloud computing is really becoming popular worldwide as it offers innumerable benefits to the clients. Most of the companies have realized the importance of this technology as it offers plenty of storage options. This type of a storage process permits all the organizations to take benefit without actually having to reimburse any additional costs that are usually connected with such type of storage resources. The characteristics and benefits of Cloud computing are as follows.

On-demand self-service: A consumer can unilaterally provision computing capabilities, such as server time and network storage, as needed automatically without requiring human interaction with each service provider.

Broad network access: Capabilities are available over the network and accessed through standard mechanisms that promote use by heterogeneous thin or thick client platforms (e.g., mobile phones, tablets, laptops and workstations).

Resource pooling: The provider's computing resources are pooled to serve multiple consumers using a multi-tenant model, with different physical and virtual resources dynamically assigned and reassigned according to consumer demand. There is a sense of location independence in that the customer generally has no control or knowledge over the exact location of the provided resources but may be able to specify location at a higher level of abstraction (e.g., country, state, or datacenter). Examples of resources include storage, processing, memory and network bandwidth.

Rapid elasticity: Capabilities can be elastically provisioned and released, in some cases automatically, to scale rapidly outward and inward commensurate with demand. To the consumer, the capabilities available for provisioning often appear to be unlimited and can be appropriated in any quantity at any time.

Measured service: Cloud systems automatically control and optimize resource use by leveraging a metering capability at some level of abstraction appropriate to the type of service (e.g., storage, processing, bandwidth and active user accounts). Resource usage can be monitored, controlled and reported, providing transparency for both the provider and consumer of the utilized service.

1.5   Cloud Computing Platforms and Technologies

Those who develop applications for Cloud computing resources in both public and private may have to make a decision on which specific Cloud computing platform to use. The wrong choice could negatively impact everyone involved, so it is important to look at the choices carefully and consider short and long term issues in our decision. The key Cloud architectures available to make this decisions are: (i) Amazon's Elastic Compute Cloud or EC2, (ii) IBM Computing or Blue Cloud, (iii) Microsoft's Azure Cloud computing, (iv) Sun Cloud, (v) Salesforce.com's Force.com Cloud and (vi) Google's AppEngine Cloud.

In addition, there are a number of open source Cloud computing tools, and several of these work together. Here, we describe five Cloud management platform (CMP) solutions, two commercial Cloud services and three open-source offerings. The open-source solutions: Apache CloudStack, Eucalyptus Cloud computing software and the OpenStack platform, typically provide a low-cost point of entry for the software and the prospect of application portability, but require a significant amount of in-house development. Commercial vendors Microsoft and VMware offer commercial off-the-shelf capabilities, and are typically higher cost than open-source offerings. Choosing the appropriate CMP for our Cloud environment depends on virtualization environment, the scope of Cloud strategy and business requirements, availability of skilled resources and the budget. Let us discuss some of them.

Apache CloudStack software is a top-level project of the Apache software foundation and calls itself a "turnkey" solution. CloudStack software provides an open and flexible Cloud orchestration platform for private and public Clouds. It offers self-service IaaS capabilities and features such as compute orchestration, user and account management, native API and Amazon Web Services (AWS) API translator. Apps written for CloudStack can run in AWS, alongside resource accounting of network, compute, storage resources, multi-tenancy and account separation.

Eucalyptus Systems is an open-source provider of Cloud management software with strong technical ties to Amazon Web Services. One of the advantages to deploying the Eucalyptus Cloud platform is the ability for a company to move seamlessly from a private Cloud to a hybrid model by bursting into the Amazon public Cloud as needed. Eucalyptus software supports industry-standard AWS APIs, including Amazon Elastic Compute Cloud (Amazon EC2), Amazon Simple Storage Service (Amazon S3), Amazon Elastic Block Store (Amazon EBS) and Amazon Identity and Access Management (Amazon IAM). It supports three hypervisors: VMware ESXi with vSphere technology, KVM software and the Xen Cloud Platform (XCP).

Microsoft Hyper-V Software and Microsoft System Center is referred to as Microsoft Cloud OS, a set of technologies, tools and processes built on the Windows Server operating system with Hyper-V software, the Microsoft System Center and the Windows Azure platform. Together, these technologies provide a consistent platform for infrastructure, applications and data.

OpenStack Cloud Software was cofounded by Rackspace and NASA in 2010 and is currently available under the Apache 2.0 license. Growth in the use of the OpenStack platform has been rapid, with dozens of companies, many of them well known, such as AT&T, HP and IBM, signing on to use OpenStack as the base for their private Cloud offerings. This gives IT departments two options for deploying OpenStack for private Cloud either as a free software download with in-house deployment or from a vendor.

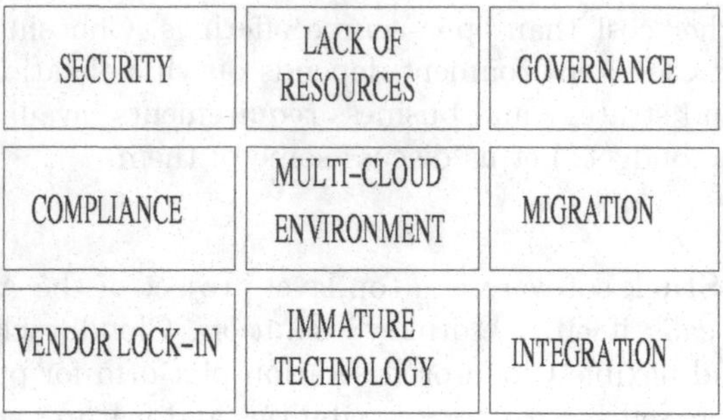

SECURITY	LACK OF RESOURCES	GOVERNANCE
COMPLIANCE	MULTI-CLOUD ENVIRONMENT	MIGRATION
VENDOR LOCK-IN	IMMATURE TECHNOLOGY	INTEGRATION

FIGURE 1.3
Challenges of Cloud computing

VMware vCloud Director is a comprehensive, integrated Cloud platform that includes all the elements to build Cloud environments and operationalize VMware vSphere virtualized environments. VMware vCenter Server manages the compute, storage, and networking resources, and VMware vCloud Director ties all the pieces of the Cloud together so that we can deploy a secure, multitenant Cloud (where multiple independent instances of one or multiple applications operate in a shared environment) using the resources from VMware vSphere environments.

Whether it's on-premises or in the Cloud, an application platform can be thought of as comprising three parts: (i) A foundation: Nearly every application uses some platform software on the machine it runs on. This typically includes various support functions, such as standard libraries and storage, and a base operating system, (ii) A group of infrastructure services: In a modern distributed environment, applications frequently use basic services provided on other computers. It is common to provide remote storage, for example, integration services, an identity service, and more and (iii) A set of application services: As more and more applications become service-oriented, the functions they offer become accessible to new applications. Even though these applications exist primarily to provide services to end users, this also makes them part of the application platform. It might seem odd to think of other applications as part of the platform, but in a service-oriented world, they certainly are.

Open Research Challenges: Cloud computing introduces many challenges for system and application developers, engineers, system administrators and service providers (see Figure 1.3). These include the following:

Security : Since the advent of the public Cloud, enterprises have worried about potential security risks, and that has not changed. A 2018 Crowd Research Partners survey found that 90 percent of security professionals are concerned about Cloud security. More specifically, they have fears about data loss and leakage (67 percent), data privacy (61 percent) and breaches of confidentiality (53 percent).

Lack of resources: Many companies are hoping to overcome this challenge by hiring more workers with Cloud computing certifications or skills. Experts also recommend providing training to existing staff to help get them up to speed with the technology.

Governance: Governance and control were fourth in the list of Cloud computing challenges in the RightScale survey with 71 percent of respondents calling it a challenge, including 25 percent who see it as a significant challenge. Experts say organizations can alleviate some of these Cloud computing management issues by following best practices, including establishing and enforcing standards and policies.

Compliance: The recent flurry of activity surrounding the EU General Data Protection Regulation (GDPR) has returned compliance to the forefront for many enterprise IT teams. Among those surveyed by RightScale, 68 percent cited compliance as a top Cloud computing challenge, and 21 percent called it a significant challenge.

Multi-Cloud environment: Most organizations are not using just one Cloud. According to the RightScale findings, 81 percent of enterprises are pursuing a multi-Cloud strategy, and 51 percent have a hybrid Cloud strategy (public and private Clouds integrated together).

Migration: While launching a new application in the Cloud is a fairly straightforward process, moving an existing application to a Cloud computing environment is not an easy task.

Vendor lock-in: Currently, a few vendors, namely Amazon Web Services, Microsoft Azure, Google Cloud Platform and IBM Cloud, dominate the public Cloud market. For both analysts and enterprise IT leaders, this raises the specter of vendor lock-in.

Immature technology: Many Cloud computing services are on the cutting edge of technologies like artificial intelligence, machine learning, augmented reality, virtual reality and advanced big data analytics. The potential downside to access to this new and exciting technology is that the services do not always live up to enterprise expectations in terms of performance, usability and reliability.

Integration: Lastly, many organizations, particularly those with hybrid Cloud environments report challenges related to getting their public Cloud and on-premise tools and applications to work together.

This challenge, like the others mentioned, is unlikely to disappear any time in the near future. Integrating legacy systems and new Cloud-based applications requires time, skill and resources. But many organizations are finding that the benefits of Cloud computing outweigh the potential downside of the technology.

Summary

Following a brief history of Cloud computing and a discussion of business drivers and technology innovations, basic terminology and concepts are introduced in this chapter, along with descriptions of common benefits and challenges of Cloud computing adoption. Cloud delivery and deployment models are discussed in detail, followed by sections that establish common Cloud characteristics and roles and boundaries. Tools and technologies in Cloud computing are also presented. The chapter concludes with the open challenges in the field of Cloud computing.

Keywords

Cloud Computing	Private Cloud	IaaS
Virtual Machines	Hybrid Cloud	PaaS
Public Cloud	CloudSim	SaaS

Objective type questions

1. _____ computing refers to applications and services that run on a distributed network using virtualized resources.

   (a) Distributed   (b) Cloud   (c) Soft   (d) Parallel

2. Point out the wrong statement:

   (a) The massive scale of Cloud computing systems was enabled by the popularization of the Internet

   (b) Soft computing represents a real paradigm shift in the way in which systems are deployed

   (c) Cloud computing makes the long-held dream of utility computing possible with a pay-as-you-go, infinitely scalable, universally available system

   (d) None

3. _____ as a utility is a dream that dates from the beginning of the computing industry itself.

   (a) Model   (b) Computing

   (c) Software   (d) All of the mentioned above

4. Which of the following is essential concept related to Cloud ?

   (a) Reliability   (b) Productivity

   (c) Abstraction   (d) All of the mentioned above

5. Point out the wrong statement:

   (a) All applications benefit from deployment in the Cloud   (b) With Cloud computing, you can start very small and become big very fast

   (c) Cloud computing is revolutionary, even if the technology it is built on is evolutionary   (d) None

6. Which of the following Cloud concept is related to pooling and sharing of resources ?

   (a) Polymorphism   (b) Abstraction

   (c) Virtualization   (d) None of the mentioned

7. _____ has many of the characteristics of what is now being called Cloud computing.

   (a) Internet   (b) Softwares

   (c) Web Service   (d) All of the mentioned

8. Which of the following can be identified as Cloud ?

   (a) Web Applications   (b) Intranet

   (c) Hadoop   (d) All of the mentioned

9. Cloud computing is an abstraction based on the notion of pooling physical resources and presenting them as a _____ resource.

   (a) real   (b) virtual

   (c) Cloud   (d) None of the mentioned

10. Which of the following is Cloud Platform by Amazon ?

    (a) Azure   (b) AWS

    (c) Cloudera   (d) All of the mentioned

11. _____ describes a Cloud service that can only be accessed by a limited amount of people.

    (a) Data center   (b) Private Cloud

    (c) Virtualization   (d) Public Cloud

12. _____ describes a distribution model in which applications are hosted by a service provider and made available to users.

    (a) Infrastructure-as-a-Service   (b) Platform-as-a-Service (PaaS)
        (IaaS)

    (c) Software-as-a-Service (SaaS)   (d) Cloud service

13. Access to a Cloud environment always costs more money compared to a traditional desktop environment.

    (a) True   (b) False

14. _____ is the feature of Cloud computing that allows the service to change in size or volume in order to meet a user's needs.

    (a) Scalability   (b) Virtualization

    (c) Security   (d) Cost-savings

15. A Cloud environment can be accessed from anywhere in the world as long as the user has access to the Internet.

    (a) True   (b) False

Objective type questions -answer

1: b   2: b   3: b   4: c   5: a   6: c   7: a   8: c   9: b   10: c   11: b   12: c   13: b   14: a   15: b

Review questions

1. What are the advantages of using Cloud computing ?
2. Mention the platforms which are used for large scale Cloud computing.
3. Explain different models for deployment in Cloud computing.
4. Mention platforms which are used for large scale Cloud computing.

5. What is the difference in Cloud computing and computing for mobiles ?
6. How user can gain from utility computing ?
7. What are the security aspects provided with Cloud ?
8. List out different layers which define Cloud architecture.
9. What does "EUCALYPTUS" stands for ?
10. What is the requirement of virtualization platform in Cloud ?
11. Mention the name of some large Cloud providers and databases.
12. Explain the difference between Cloud and traditional datacenters.

Critical thinking questions

1. An organization debating whether to install a private Cloud or to use a public Cloud, e.g., the AWS, for its computational and storage needs, asks your advice. What information will you require to base your recommendation on, and how will you use each one of the following items: (a) the description of the algorithms and the type of the applications the organization will run; (b) the system software used by these applications; (c) the resources needed by each application; (d) the size of the user population; (e) the relative experience of the user population; (d) the costs involved.

2. Non-linear algorithms do not obey the rules of scaled speed-up. For example, it was shown that when the concurrency of an $O(N^3)$ algorithm doubles, the problem size increases only by slightly more than 25%. Read this research paper and explain the result: L. Snyder, "Type architectures, shared memory, and the corolary of modest potential", Ann. Rev. Comp. Sci. 1, pp. 289-317, 1986.

3. Given a system of four concurrent threads $t1$, $t2$, $t3$ and $t4$, we take a snapshot of the consistent state of the system after 3, 2, 4 and 3 events in each thread, respectively; all but the second event in each thread are local events. The only communication event in thread $t14$ is to send a message to $t4$ and the only communication event in thread $t3$ is to send a message to $t2$. Draw a space-time diagram showing the consistent cut. How many messages are exchanged to obtain the snapshot in this case? The snapshot protocol allows the application developers to create a checkpoint. An examination of the checkpoint data shows that an error has occurred and it is decided to trace the execution. How many potential execution paths must be examined to debug the system?

4. Several desirable properties of a large-scale distributed system includes transparency of access, location, concurrency, replication, failure, migration, performance and scaling. Analyze how each one of these properties applies to AWS.

Bibliography

[1] Rajkumar Buyya and Rajiv Ranjan, "Federated Resource Management in Grid and Cloud Computing Systems", *Elsevier Journal of Future Generation Computer System*, Vol. 26, No. 5, 2013, pp. 1189-1191.

[2] "Cloud Computing", *Available online at http://Cloudcomputing.syscon.com/node/612375*, Accessed on 12.07.2012.

[3] Dai. W, Rubin, "Service-Oriented Knowledge Management Platform", *Proceedings of the 13th IEEE International Conference on Information Reuse and Integration*, Las Vegas, USA, 2013, pp. 255-259.

[4] "Cloud Project", *Available online at http://www.ibm.com/developerworks/web/library/wa-Cloudgrid*, Accessed on 10.07.2018.

[5] B. Urgaonkar, P. Shenoy, A. Chandra, P. Goyal, and T. Wood, "Agile Dynamic Provisioning of Multi-tier Internet Applications", *ACM Transactions on Autonomous and Adaptive Systems*, Vol. 5, No. 5, 2010, pp. 139-148.

[6] L.M. Vaquero, L.R. Merino, J. Caceres, and M. Lindner, "A Break in the Clouds: Towards a Cloud Definition", *ACM SIGCOMM Computer Communication Review*, Vol. 39, No. 1, 2009, pp. 67-72.

[7] M. Gupta and S. Singh, "Greening of the Internet", *Proceedings of ACM International Conference on Applications, Technologies, Architectures, and Protocols for Computer Communication*, London, UK, 2014, pp. 29-36.

[8] L. Chiaraviglio and I. Matta, "Co-operative Green Routing with Energy Efficient Servers", *Proceedings of ACM International Conference on Energy Efficient Computing and Networking*, New York, USA, 2010, pp. 191-194.

[9] Flavio Lombardi and Roberto Di Pietro, "Secure Virtualization for Cloud Computing", *Elsevier Journal of Network and Computer Applications*, Vol. 41, No. 1, 2011, pp. 45-52.

[10] Ibrahim. S and Shl. X, "Evaluating Mapreduce on Virtual Machines: The Hadoop Case", *Proceedings of the 1st ACM International Conference on Cloud Computing*, Beijing, China, 2009, pp. 345-350.

[11] Arun Haider, Richard Potter, and Akihiro Nakao, "Challenges in Resource Allocation in Network Virtualization", *Proceedings of the ITC Specialist Seminar on Network Virtualization*, Hoian, Vietnam, 2009, pp. 34-41.

[12] P. Ruth, P. McGachey, and D. Xu, "Viocluster: Virtualization for Dynamic Computational Domains", *Proceedings of the IEEE International Conference on Cluster Computing*, London, UK, 2015, pp. 19-26.

[13] K. Keahey, I. Foster, T. Freeman, and X. Zhang, "Achieving Quality of Service and Quality of Life in the Grid, Scientific Programming", *Journal of Future Generation Computer System*, Vol. 13, No. 4, 2011, pp. 265-275.

[14] "OpenNebula project", *Available online at http://www.opennebula.org*, Accessed on 09.05.2017.

2

Distributed and Parallel Computing

Learning Objectives

After reading this chapter, you will be able to

- Describe the fundamentals of distributed and parallel computing and their properties

- Differentiate between distributed and parallel computing and understand their pros and cons

- Analyse performance considerations in distributed and parallel computing

Parallel and Distributed Computing has made it possible to simulate large infrastructures like Telecom networks, air traffic etc. in an easy and effective way. In this chapter, we discuss the problem of coordinating multiple computers and processors. First, we will look at distributed systems. These are interconnected groups of independent computers that need to communicate with each other to execute several jobs. Coordination is required to provide a service, share data, or even store data sets that are too large to fit on a single machine. Then, we shall look at different roles computers can play in distributed systems and learn about the kinds of information that computers need to exchange in order to work together.

Next, we will consider concurrent computation, also known as parallel computation. Concurrent computation is when a single program is executed by multiple processors with a shared memory, all working together in parallel in order to get work done faster. Concurrency introduces new challenges, and so we will see techniques to manage the complexity of concurrent programs.

Preliminaries

The following are some of the terms and terminologies defined here for easier understanding in the remaining part of this chapter.

Middleware: It is a software that acts as a bridge between an operating system or database and applications, especially on a network.

Message passing: It is a type of communication between processes. Message passing is a form of communication used in parallel programming and object-oriented programming. Communications are completed by the sending of messages (functions, signals and data packets) to recipients.

EJB container: Enterprise Java Beans (EJB components) are Java programming language server components that contain business logic. The EJB container provides local and remote access to enterprise beans.

JMS Queue: A staging area that contains messages that have been sent and are waiting to be read (by only one consumer). As the name queue suggests, the messages are delivered in the order sent. A JMS queue guarantees that each message is processed only once.

XML: Extensible Markup Language (XML) is a markup language that defines a set of rules for encoding documents in a format that is both human-readable and machine-readable.

WS-I: The Web Services Interoperability (WS-I) organization is an open industry organization chartered to establish best practices for web services interoperability, for selected groups of web services standards, across platforms, operating systems and programming languages.

SOAP: Simple Object Access Protocol (SOAP) is a messaging protocol specification for exchanging structured information in the implementation of web services in computer networks.

REST: Representational state transfer (REST) is a software architectural style that defines a set of constraints to be used for creating Web services. Web services that conform to the REST architectural style, called RESTful Web services, provide interoperability between computer systems on the Internet.

Hyperthreading: It is a technology used by microprocessors that allows a single microprocessor to act like two separate processors to the operating system and the application programs that use it.

Remote Method Invocation (RMI): It is an API which allows an object to invoke a method on an object that exists in another address space, which could be on the same machine or on a remote machine.

Parallelism: Parallelism is basically a type of computation in which many computations or operations are carried out in parallel. This is done to achieve the speed up in computation.

2.1   Distributed Computing

A distributed system is a network of autonomous computers that communicate with each other in order to achieve a goal. The computers in a distributed system are independent and do not physically share memory or processors. Figure 2.1 shows the distributed system: each computer has its own local memory, and information can be exchanged only by passing messages from one node to another by using the available communication links.

Computers in a distributed system can have different roles. A computer's role depends on the goal of the system and the computer's own hardware and software properties. There are two predominant ways of organizing computers in a distributed system. The first is the client-server architecture, and the second is the peer-to-peer architecture.

Client/Server Systems

The client-server model of communication can be traced back to the introduction of UNIX in the 1970's, but perhaps the most influential use of the model is the modern World Wide Web. An example of a client-server interaction is reading the New York Times online. When the web server at www.nytimes.com is contacted by a web browsing client (like Firefox), server's job is to send back the HTML of the New York Times main page. This could involve calculating personalized content based on user account information sent by the client, and fetching appropriate advertisements. The job of the web browsing client is to render the HTML code sent by the server. This means displaying the images, arranging the content visually, showing different colors, fonts and shapes and allowing users to interact with the rendered web page.

The client-server architecture is a way to dispense a service from a central source. There is a single server that provides a service, and multiple clients that communicate with the server to consume its products. In this architecture, clients and servers have different jobs. The server's job is to respond to service requests from clients, while a client's job is to use the data provided in response to perform some specific task.

The concepts of client and server are powerful functional abstractions. A server is simply a unit that provides a service, possibly to multiple clients simultaneously, and a client is a unit that consumes the service. The clients do not need to know the details of how the service is provided, or how the data they are receiving is stored or calculated, and the server does not need

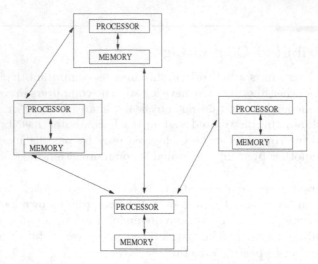

FIGURE 2.1
Distributed System

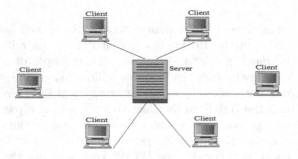

FIGURE 2.2
Client Systems

to know how the data is going to be used. A drawback of client-server systems is that the server is a single point of failure (Fig. 2.2). It is the only component with the ability to dispense the service. There can be any number of clients, which are interchangeable and can come and go as necessary. If the server goes down, the system stops working. Thus, the functional abstraction created by the client-server architecture also makes it vulnerable to failure. Another drawback of client-server systems is that resources become scarce if there are too many clients. Clients increase the demand on the system without contributing any computing resources. Client-server systems cannot shrink and grow with changing demand.

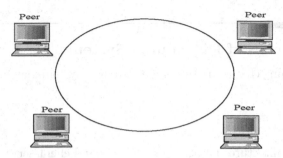

FIGURE 2.3
Peer-to-Peer Systems

Peer-to-peer Systems

The client-server model is appropriate for service-oriented situations. However, there are other computational goals for which a more equal division of jobs is a better choice. The term peer-to-peer is used to describe distributed systems in which jobs are divided among all the components of the system. All the computers send and receive data, and they all contribute some processing power and memory. As distributed system increases in size, its capacity of computational resources increases. In a peer-to-peer system, all components of the system contribute to some processing power and memory for a distributed computation (Fig. 2.3).

Division of jobs among all participants is the identifying characteristic of a peer-to-peer system. This means that peers need to be able to communicate with each other reliably. In order to make sure that messages reach their intended destinations, peer-to-peer systems need to have an organized network structure. The components in these systems cooperate to maintain enough information about the locations of other components to send messages to intended destinations.

The most common applications of peer-to-peer systems are data transfer and data storage. For data transfer, each computer in the system contributes to send data over the network. For data storage, the data set may be too large to fit on any single computer, or too valuable to store on just a single computer. Each computer stores a small portion of the data, and there may be multiple copies of the same data spread over different computers. When a computer fails, the data that was on it can be restored from other copies and put back when a replacement arrives.

Skype, the voice and video-chat service, is an example of a data transfer application with a peer-to-peer architecture. When two people on different computers are having a Skype conversation, their communications are broken up into packets of 1s and 0s and transmitted through a peer-to-peer network.

2.2   Properties of Distributed Systems

The distinguishing characteristics of a distributed system may be summarized as follows.

Modularity

The two architectures namely, peer-to-peer and client-server are designed to enforce modularity. Modularity is the idea that the components of a system should be black boxes with respect to each other. It should not matter how a component implements its behavior as long as it upholds an interface: a specification for what outputs will result from inputs.

Modularity gives a system many advantages, and is a property of thoughtful system design. First, a modular system is easy to understand. This makes it easier to change and expand. Second, if something goes wrong with the system, only the defective components need to be replaced. Third, bugs or malfunctions are easy to localize. If the output of a component doesn't match the specifications of its interface, even though the inputs are correct, then that component is the source of the malfunction.

Message Passing

In distributed systems, components communicate with each other using message passing. A message has three essential parts: the sender, the recipient, and the content. The sender needs to be specified so that the recipient knows the origin of the message. The recipient needs to be specified so that any computers who are helping send the message know where to direct it. Depending on the function of the overall system, the content of the message can be a piece of data, a signal, or instructions for the remote computer to evaluate a function with some arguments.

In a distributed system, messages may need to be sent over a network, and may need to hold many different kinds of signals as "data," so they are not always encoded as strings. In both cases, however, messages serve the same function. Different components (dispatch functions or computers) exchange them in order to achieve a goal that requires coordinating multiple modular components.

At a high level, message contents can be complex data structures, but at a low level, messages are simply streams of 1s and 0s sent over a network. In order to be usable, all messages sent over a network must be formatted according to a consistent message protocol. A message protocol is a set of rules for encoding and decoding messages. Many message protocols specify that a message conform to a particular format, in which certain bits have a consistent meaning. A fixed format implies fixed encoding and decoding rules to generate and read that format. All the components in the distributed system must understand the protocol in order to communicate with each other. That way, they know which part of the message corresponds to which information.

2.3 Performance Consideration in Distributed Computing

Distribution and communication between applications and services is a central concept in modern application architectures. In order to profit from distribution, some basic principles must be borne in mind, otherwise one can easily run into performance and scalability problems. During development, these problems often do not surface. Later, while load testing or production, the chosen software architecture may not support the required performance and scalability requirements. In this section, we will look at major points to keep in mind for building distributed applications. Some of the important performance consideration in distributed computing are: (i) High-level architecture of remoting protocols, (ii) Sychronous and asynchronous Simple Object Access Protocol (SOAP) architectures. They are discussed as follows.

2.3.1 High-Level Architecture of Remoting Protocols

The beauty of remoting technologies is that there are many to choose from. Java offers a huge variety of possibilities and technologies to implement distributed applications. The selection of a remoting technology already significantly influences the architecture and also performance and scalability of an application. The "oldest" and assumedly most widely used remoting protocol is Remote Method Invocation (RMI) (Fig. 2.4). RMI is the standard protocol for Java Enterprise Edition (JEE) applications. As the name implies, it is designed for invoking methods of objects hosted in other Java Virtual Machines (JVMs). Objects are exposed at the server side and can then be invoked from clients via proxies. The same server object is used by multiple threads. The thread pool is managed by the RMI infrastructure.

The steps in RMI-based application are shown in Figure 2.4. It involves: (1) Design of the interface for the service, (2) Implement the methods specified in the interface, (3) Generate the stub and the skeleton, (4) Register the service by name and location, and (5) Use the service in an application.

As references to the server side have to be managed as well, the RMI infrastructure also provides a specific garbage collector to manage remote

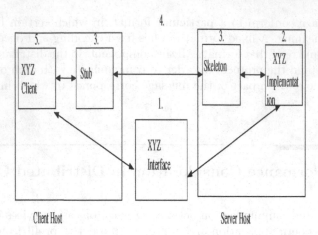

FIGURE 2.4
RMI-based distributed system

references. The Distributed Garbage Collector (DGC) itself makes use of the RMI protocol for managing the server-side objects' lifecycle. Besides the strong coupling of the client and the server RMI comes with a number of further implications. RMI only supports synchronous communication with all the disadvantages discussed above. Developer and system architects can influence the configuration parameters of the infrastructure to optimize performance.

Java Messaging Service (JMS) is the second most commonly used protocol in the Java platform enterprise edition (JEE) space (see Fig. 2.5). In contrast to RMI, JMS is an asynchronous protocol. The communication is based on queues where listeners are used to react on messages. JMS is not a classical remote procedure call protocol but still fits well for service-to-service interactions. In many Enterprise Service Bus (ESB) implementations which often act as the central point in SOAs, a JMS-based middleware is used to exchange messages between services. Due to its asynchronicity, the typical problems of synchronous processing can be avoided. In many systems, a central aspect for scalability is to free up resources (like threads) very fast. Asynchronous processing approaches are in many cases the only suitable way to do that.

Main Components in JMS

1. Publisher: Responsible to publish message to Message Queue (MQ) server

2. MQ Server/Message Broker: Holds the messages in MQ server

3. Subscriber: Responsible for performing the task on the basic of message posted by publisher.

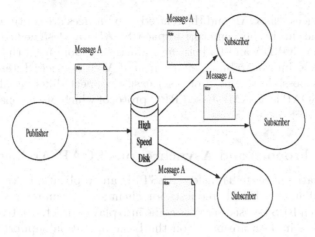

FIGURE 2.5
JMS Architecture

Type of Communication Supported or JMS Models

1. Point to Point: One Message is consumed by one client (One to One)

2. Publish-Subscribe: One Message is consumed by multiple-clients (One to Many)

3. JMS API: It is used for developing message-oriented-middleware (MOM) based application in Java

4. Message Driven Bean (MDB): A message driven bean is a stateless, server-side, transaction-aware component that is driven by a Java message (javax.jms.message). It is invoked by the EJB Container when a message is received from a JMS Queue or Topic. It acts as a simple message listener.

5. Spring Framework: It is an application framework and inversion of control container for the Java platform. The framework's core features can be used by any Java application, but there are extensions for building web applications on top of the JEE platform. Although the framework does not impose any specific programming model, it has become popular in the Java community as an addition to, or even replacement for the EJB model. The Spring Framework is open source.

JMS offers a number of different transport formats. XML is the most commonly used message format, but binary formats would also be possible. The design of the message structure must be a central part in the application architecture as it directly influences performance and scalability.

Web Services via SOAP and the related web services are continuously gaining importance in the Java enterprise space. SOAP was designed to provide an alternative to CORBA and had strong industry support from the beginning. Because of the interoperability effort around Web Services Interoperability (WS-I), it is possible to (more or less) easily connect different platforms to each other. SOAP is an XML-based RPC protocol which often consumes more bandwidth.

2.3.2   Sychronous and Asynchronous SOAP Architecture

A Representational State Transfer (REST) is an Application Program Interface (API) that uses HTTP requests for client-server communication. More and more often REST-based services come into play as an alternative to SOAP. REST Services in Java are based on the basic operation support by HTTP. REST however is not designed to be used as an Remote Procedure Call (RPC) protocol. It is rather resource oriented and designed for access and manipulation of web resources. Both protocols support synchronous communication. This is also mandated by the underlying HTTP protocol. The WS-Addressing extension for SOAP however also allows the implementation of asynchronous services. A big advantage of REST is the ability to easily implement caching by using HTTP proxies. REST is relying on the mechanisms which are anyway available via the underlying HTTP protocol.

2.4   Parallel Computing

Parallel computing is a type of computation in which many calculations are carried out simultaneously, operating on the principle that large problems can often be divided into smaller ones, which are then solved at the same time. There are several different forms of parallel computing: bit-level, instruction-level, data, and task parallelism. Parallelism has been employed for many years, mainly in high-performance computing, but interest in it has grown lately due to the physical constraints preventing frequency scaling, a technique in computer architecture whereby the frequency of a microprocessor can be automatically adjusted "on the fly" depending on the actual needs, to conserve power and reduce the amount of heat generated by the chip. As power consumption (and consequently heat generation) by computers has become a concern in recent years, parallel computing has become the dominant paradigm in computer architecture, mainly in the form of multi-core processors.

Parallel computing is closely related to concurrent computing: they are frequently used together, and often confused, though the two are distinct: it is possible to have parallelism without concurrency (such as bit-level parallelism), and concurrency without parallelism (such as multi-tasking by time-sharing on a single-core CPU). In parallel computing, a computational task

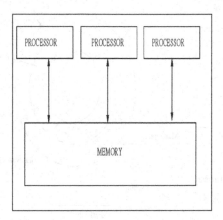

FIGURE 2.6
Parallel System

is typically broken down in several, often many, very similar subtasks that can be processed independently and whose results are combined afterward, upon completion. In contrast, in concurrent computing, the various processes often do not address related tasks; when they do, as is typical in distributed computing, the separate tasks may have a varied nature and often require some inter-process communication during execution. Figure 2.6 shows a parallel system in which each processor has a direct access to a shared memory.

Virtually all stand-alone computers today are parallel from a hardware perspective with multiple functional units, multiple execution units/cores, multiple hardware threads. Figure 2.7 shows parallel computer architecture model. Modern computers have powerful and extensive software packages. To analyze the development of the performance of computers, the basic development of hardware and software are presented as follows.

Computer Development Milestones: There are two major stages of development of computer – mechanical or electromechanical parts. Modern computers evolved after the introduction of electronic components. High mobility electrons in electronic computers replaced the operational parts in mechanical computers. For information transmission, electric signal which travels almost at the speed of a light replaced mechanical gears or levers.

Elements of Modern computers: A modern computer system consists of computer hardware, instruction sets, application programs, system software and user interface. The computing problems are categorized as numerical computing, logical reasoning and transaction processing. Some complex problems may need the combination of all the three processing modes.

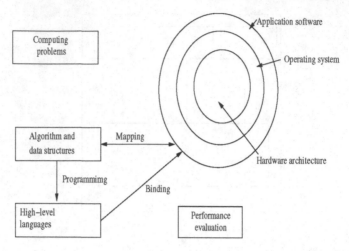

FIGURE 2.7
Parallel computer architecture model

Evolution of Computer Architecture: In last four decades, computer architecture has gone through revolutionary changes. We started with Von Neumann architecture and now we have multicomputers and multiprocessors.

Performance of a computer system: Performance of a computer system depends both on machine capability and program behavior. Machine capability can be improved with better hardware technology, advanced architectural features and efficient resource management. Program behavior is unpredictable as it is dependent on application and run time conditions.

2.5 Performance Consideration in Parallel Computing

The performance of a parallel processor depends on complex and hard to define ways on the system's architecture and degree of parallelism in programs it executes. There are several simple performance measures such as speedup and processor utilization (efficiency) which provides rough performance estimates. The system interconnection structure and main-memory architecture also have a significant impact on performance. The performance consideration in parallel computing are: (i) Parallelism degree, (ii) SpeedUp, and (iii) Efficiency. They are discussed as follows.

Parallelism Degree

Pipelining is a set of data processing elements connected in series, where the output of one element is the input of the next one. Individual or average instruction execution times are not very useful, since they do not measure the impact of what might be loosely called the system's parallelism degree n, e.g., the number of processors or the number of pipeline segments available. In theoretical studies, the time $T(n, N)$ needed to execute more complex operations such as sorting a list of N numbers, or inverting an $N*N$ matrix, is given as a performance measure for a machine of parallelism n. Such measures may be expressed in approximate form using the $\mathcal{O}$-notation $\mathcal{O}(\ f(n, N))$, where stating that $T(n, N) = \mathcal{O}(\ f(n, N))$ means that there exist constants k and $N_0 > 0$, whose exact values are not usually specified, such that $T(n, N) \leq kf(n, N)$ for all $N > N_0$. As N, the problem size or dimension increases beyond some lower limit N_0, the execution time $T(N)$ grows at a rate that is less, within a constant k, than that of the function $f(n, N)$.

The performance of an individual processor or processing elements in high performance machines may be measured by the instruction rate or instruction bandwidth b_l, using units Million Instructions Per Second (MIPS). The corresponding measures of data bandwidth or throughput b_d is typically Mflops; other multiple such as GFLOPS are also used.

In parallel computers, the interprocessor communication mechanism and the extent to which they are used, strongly influences overall system performance. For example for a system of n processors, each with performance p, maximum or potential performance turns out to be $n*p$.

Speedup

Speedup, $Sp(n)$, is defined as the ratio of the total execution time $T(l)$ on a sequential computer to the corresponding execution time $T(n)$ on the parallel computer for some class of tasks of interest (Equation 2.1).

$$Sp(n) = \frac{T(l)}{T(n)} \qquad (2.1)$$

It is usually reasonable to assume that $T(l) \leq n*T(n)$, in which case $Sp(n) \leq n$.

Efficiency

A closely related performance measure, which can be expressed as a single number (a fraction or a percentage), is the efficiency E_n, which is the speedup per degree of parallelism (Equation 2.2).

$$E(n) = \frac{S(n)}{n} \qquad (2.2)$$

$E(n)$ is also an indication of processor utilization.

In general, speed up and efficiency provide rough estimates of the performance changes that can be expected in a parallel processing system by increasing the parallelism degree n, e.g., by adding more processors. These measures should be used with caution, however, since they depend on the program being run, and can change dramatically from program to program, or from one part of the program to another.

2.6 Amdahl's Law

A program or an algorithm Q may sometimes be characterized by its degree of parallelism n_i, which is the minimum value of n for which the efficiency and speedup reach their maximum values at time i during the execution of Q. Thus n_i represents the maximum level of parallelism that can be exploited by Q at time i.

Some indication of the influence of program parallelism on the performance of a parallel computer may be seen from the following analysis. Suppose that all computations of interest on a parallel processor can be divided into two simple groups involving floating point arithmetic only: vector operations employing vector operands of some fixed length N, and scalar operations where all operands are scalars($N=1$).

Let $1 - f$ be the fraction of all floating point operations that are executed as scalar operations, and let f be the fractions executed as vector operations. f is thus a measure of the degree of parallelism in the program being executed, and varies from 1, corresponding to maximum parallelism (all vector operations), to 0 (all scalar operations). Suppose that vector and scalar operations are performed at throughput rates of V and S Mflops, respectively. Let the total CPU time be t . Let a given algorithm has N flops. Then, relation of f, N, V, S, t are given by the useful formula as given in Equation 2.3.

$$t = \frac{(1-f)N}{S} + \frac{fN}{V} \qquad (2.3)$$

It can be used to evaluate the performance of parallel processors in the context of vector and scalar operations. It is found out that the parallel processor is very sensitive to the scalar operations. A very small fraction of scalar operations drops the system performance by a significant amount. Hence, it is worthwhile to devote considerable effort to "vectorize" or "parallelize" programs for such machines to eliminate scalar operations. With $1 - f$ interpreted broadly as the fraction of nonparallelizable operations or instructions. Equation 2.3 is often referred as ***Amdahl's Law*** . If the algorithm had been executed completely with the lower speed S, the CPU time would have been $t = \frac{N}{S}$, where N is the number of operations in algorithm. This implies that the relative gain in the CPU time obtained from executing the portion fN at the speed V instead of the much lower speed S is bounded by $\frac{1}{1-f}$. Thus, f must be rather close to 1 in order to benefit significantly from high computational speed.

General form of Amdahl's Law

Assume that the execution of an algorithm consists of executing the consecutive parts $A_1, A_2, ..., A_n$ such that the N_j flops of part A_j are executed at a speed of r_j Mflops. Then the overall performance r for the $N = N_1 + N_2 + ... + N_n$ flops is given by Equation 2.4.

$$r = \frac{N}{\frac{N_1}{r_1} + \frac{N_2}{r_2} + ... + \frac{N_n}{r_n}} Mflops \tag{2.4}$$

This follows directly from the observation that the CPU time t_j for the execution of part A_j is given as in Equation 2.4.

From Amdahl's law we can study the effect of different speeds for separate parts of a computational task. It is seen that the improvements in the execution of computations usually are obtained only after the startup time. This fact implies that the computational speed depends strongly on the amount (and type) of the work.

2.7 Types of Parallelism

Parallel processing is a vital part of any high performance computing model. It involves the utilization of large amounts of computing resources to complete a complex task or problem. The resources specific to parallel processing are CPU and memory. Originally confined to use in scientific applications, parallel processing has quickly made inroads into commercial and business applications that need high performance computing facilities like data mining, decision support and risk management applications.

Parallel execution or processing involves the division of a task into several smaller tasks and making the system work on each of those smaller tasks in parallel. In simple terms, if multiple processors engage a computing task, it is generally executed faster. Parallel processing thereby improves response time and increases throughput by utilizing all of the available computing resources.

Parallel execution helps systems scale performance by making optimal use of the hardware resources. In a parallel processing system, multiple processes may reside on a single computer or may be spread across several computers or nodes, as in an Oracle10g RAC cluster.

Some basic requirements for achieving parallel execution and better performance are: (i) Computer system/servers with built in multiple processors and better message facilitation among processors, (ii) Operating system capable of managing the multiple processors, and (iii) Clustered nodes with application software, such as Oracle Real Application Clusters (RAC), that can perform parallel computations across several nodes.

There are many distinct benefits and advantages of utilizing parallel execution, such as better response times, higher throughput, and better price/performance. There are many types of parallelism. Some of the types are Bit-level parallelism, Task-level parallelism, and Instruction-level parallelism. They are explained as follows.

Bit-level parallelism

Increasing the word size reduces the number of instructions the processor must execute to perform an operation on variables whose sizes are greater than the length of the word. For example, in case of 8-bit processor adding two 16-bit integers, the processor must first add the 8 lower-order bits from each integer using the standard addition instruction, then add the 8 higher-order bits using an add-with-carry instruction and the carry bit from the lower order addition; thus, an 8-bit processor requires two instructions to complete a single operation, where a 16-bit processor would be able to complete the operation with a single instruction.

Historically, 4-bit microprocessors were replaced with 8-bit, then 16-bit, then 32-bit microprocessors and so on. This trend generally came to an end with the introduction of 32-bit processors, which has been a standard in general-purpose computing for two decades. Not until recently, with the advent of x86-64 architectures, have 64-bit processors become commonplace.

Task-level parallelism

Task parallelisms is the characteristic of a parallel program that "entirely different calculations can be performed on either the same or different sets of data." This contrasts with data parallelism, where the same calculation is performed on the same or different sets of data. Task parallelism involves the decomposition of a task into sub-tasks and then allocating each sub-task to a processor for execution. The processors would then execute these sub-tasks simultaneously and often cooperatively. The task parallelism involves the parallelism of various tasks which enable communication between it. Figure 2.8 shows the task level parallelism, which is organizing a program or computing solution into a set of processes/ tasks/ threads for simultaneous execution.

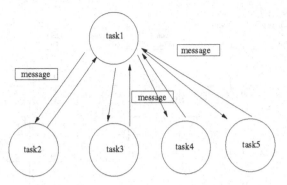

FIGURE 2.8
Task-level Parallelism

Various nodes are involved in the network and the result of the task is obtained by exchanging messages.

Instruction-level parallelism

All modern processors have multi-stage instruction pipelines. Each stage in the pipeline corresponds to a different action the processor performs on that instruction in that stage; a processor with an N-stage pipeline can have up to N different instructions at different stages of completion and thus can issue one instruction per clock cycle (IPC = 1). These processors are known as scalar processors. The canonical example of a pipelined processor is a Reduced Instruction Set Computer (RISC) processor, with five stages: instruction fetch (IF), instruction decode (ID), execute (EX), memory access (MEM), and register write back (WB). Scoreboarding and the Tomasulo algorithm (which is similar to scoreboarding but makes use of register renaming) are two of the most common techniques for implementing out-of-order execution and instruction-level parallelism. A five stage pipeline is shown in Fig. 2.9. The input of each stage is a function of time. Nine clock cycles are illustrated. Six steps are involved in execution of an instruction by CPU. However, not all of them are required for all instructions, (i) Fetch instruction, (ii) Decode information, (iii) Perform Execute/ ALU operation, (iv) Access memory, and (v) Write Back the result to the memory.

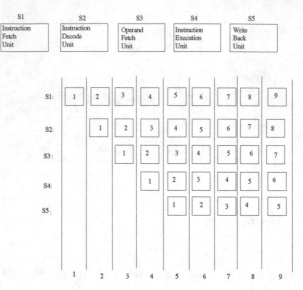

FIGURE 2.9
Instruction level Parallelism

2.8 Flynn's Classical Taxonomy

There are different ways to classify parallel computers. One of the more widely used classifications, in use since 1966, is called Flynn's Taxonomy. Flynn's taxonomy distinguishes multi-processor computer architectures according to how they can be classified along the two independent dimensions of Instruction Stream and Data Stream. Each of these dimensions can have only one of two possible states: Single or Multiple. The following classification of parallel computers have been identified: (1) Classification based on the instruction and data streams, (2) Classification based on the structure of computers, (3) Classification based on how the memory is accessed, and (4) Classification based on grain size.

The brief explaination about the four possible classifications according to Flynn are explained as follows.

Single Instruction, Single Data (SISD): The features of SISD are as follows.

- A serial (non-parallel) computer

- Single Instruction: Only one instruction stream is being acted on by the CPU during any one clock cycle

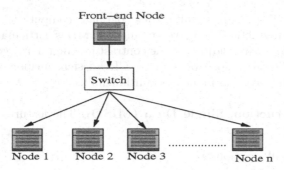

FIGURE 2.10
SIMD architecture

- Single Data: Only one data stream is being used as input during any one clock cycle

- Deterministic execution

- Examples: older generation mainframes, minicomputers, workstations and single processor/core PCs.

Single Instruction, Multiple Data (SIMD): The features of SIMD are as follows.

- A type of parallel computer

- Single Instruction: All processing units execute the same instruction at any given clock cycle

- Multiple Data: Each processing unit can operate on a different data element

- Best suited for specialized problems characterized by a high degree of regularity, such as graphics/image processing.

- Synchronous (lockstep) and deterministic execution

- Two varieties: Processor Arrays and Vector Pipelines

- Examples: Processor Arrays: Thinking Machines CM-2, MasPar MP-1 & MP-2, ILLIAC IV etc.

- Most modern computers, particularly those with graphics processor units (GPUs) employ SIMD instructions and execution units.

Figure 2.10 shows SIMD architecture model. It describes computers with multiple processing units (PU) that perform the same operation on multiple data points simultaneously. Such machines exploit data level parallelism, but not concurrency: there are simultaneous (parallel) computations, but only a single process (instruction) at a given moment. SIMD is particularly applicable to common tasks such as adjusting the contrast in a digital image or adjusting the volume of digital audio. Most modern CPU designs include SIMD instructions to improve the performance of multimedia use.

Multiple Instruction, Single Data (MISD): The features of MISD are as follows.

- A type of parallel computer

- Multiple Instruction: Each processing unit operates on the data independently via separate instruction streams.

- Single Data: A single data stream is fed into multiple processing units.

- Few (if any) actual examples of this class of parallel computer have ever existed.

- Some conceivable uses might be (i) multiple frequency filters operating on a single signal stream, and (ii) multiple cryptography algorithms attempting to crack a single coded message.

Multiple Instruction, Multiple Data (MIMD): The features of MIMD are as follows.

- A type of parallel computer

- Multiple Instruction: Every processor may be executing a different instruction stream

- Multiple Data: Every processor may be working with a different data stream

- Execution can be synchronous or asynchronous, deterministic or non-deterministic

- Currently, the most common type of parallel computer – most modern supercomputers fall into this category.

- Examples: most current supercomputers, networked parallel computer clusters and "grids", multi-processor SMP computers, multi-core PCs.

2.9   Classes of Parallel Computers

Parallel computers can be roughly classified according to the level at which the hardware supports parallelism. This classification is broadly analogous to the

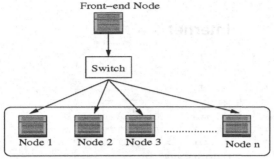

FIGURE 2.11
Cluster Computing

distance between basic computing nodes. The classes are: distributed computing, cluster computing, Grid computing, Cloud computing, massively parallel
computing, specialized parallel computers, vector processors, and multi-core
computing. They are explained as follows.

Distributed computing

A distributed computer (also known as a distributed memory multiprocessor) is a distributed memory computer system in which the processing elements are connected by a network. Distributed computers are highly scalable.

Cluster computing

Cluster computing refers that many of the computers connected on a network and they perform like a single entity. Each computer that is connected
to the network is called a node. Cluster computing offers solutions to solve
complicated problems by providing faster computational speed, and enhanced
data integrity. The connected computers execute operations all together thus
creating the impression like a single system (Fig. 2.11).

While machines in a cluster do not have to be symmetric, load balancing is
more difficult if they are not. The most common type of cluster is the Beowulf
cluster, which is a cluster implemented on multiple identical commercial off-
the-shelf computers connected with a TCP/IP Ethernet local area network.
Beowulf technology was originally developed by Thomas Sterling and Donald
Becker. The vast majority of the TOP500 supercomputers are clusters.

Grid computing

Grid computing (Fig. 2.12) is a group of networked computers which work
together as a virtual supercomputer to perform large tasks, such as analysing

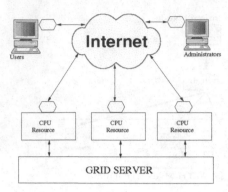

FIGURE 2.12
Grid Computing

huge sets of data or weather modeling. Through the Cloud, we can assemble and use vast computer grids for specific time periods and purposes, paying, if necessary, only for what we use to save both the time and expense of purchasing and deploying the necessary resources ourself. Also by splitting tasks over multiple machines, processing time is significantly reduced to increase efficiency and minimise wasted resources.

Unlike with parallel computing, grid computing projects typically have no time dependency associated with them. They use computers which are part of the grid only when idle and operators can perform tasks unrelated to the grid at any time. Security must be considered when using computer grids as controls on member nodes are usually very loose. Redundancy should also be built in as many computers may disconnect or fail during processing.

Cloud computing
Cloud computing is the delivery of computing services: including servers, storage, databases, networking, software, analytics, and intelligence over the Internet ("the Cloud") to offer faster innovation, flexible resources and economies of scale. We pay only for Cloud services we use, helping lower our operating costs, run our infrastructure more efficiently and scale as per our business needs.

Massively parallel computing

A massively parallel processor (MPP) is a single computer with many networked processors. MPPs have many of the same characteristics as clusters, but MPPs have specialized interconnect networks (whereas clusters use commodity hardware for networking). MPPs also tend to be larger than clusters, typically having "far more" than 100 processors. In an MPP, each CPU contains its own memory and copy of the operating system and application. Each subsystem communicates with the others via a high-speed interconnect.

Specialized parallel computers

Within parallel computing, there are specialized parallel devices that remain niche areas of interest. While not domain-specific, they tend to be applicable to only a few classes of parallel problems.

Vector processors

A vector processor is a CPU or computer system that can execute the same instruction on large sets of data. Vector processors have high-level operations that work on linear arrays of numbers or vectors. An example vector operation is A = B x C, where A, B and C are each 64-element vectors of 64-bit floating-point numbers.

Multi-core computing

A multi-core processor is a processor that includes multiple processing units (called "cores") on the same chip. This processor differs from a superscalar processor, which includes multiple execution units and can issue multiple instructions per clock cycle from one instruction stream (thread); in contrast, a multi-core processor can issue multiple instructions per clock cycle from multiple instruction streams.

Summary

This chapter covered emerging areas of computational science that pose tremendous challenge for algorithms and system development. Parallel computing has made a tremendous impact on variety of areas ranging from computational simulations for scientific and engineering applications to commercial applications in data mining and transaction processing. In a distributed system, a set of processes communicate by exchanging messages over a communication network. A distributed computation is spread over geographically distributed processes. The processes do not share a common global memory or physical global clock, to which processes have instantaneous access.

Keywords

Distributed Computing	Parallel Computing
Multiprocessors	Amdahl's Law

Objective type questions

1. In distributed system each processor has its own _____.

(a) Local memory   (b) Clock

(c) Both (a) and (b)   (d) None of the above

2. If one site fails in distributed system

(a) the remaining sites can continue   (b) all the sites will stop working
    operating

(c) directly connected sites will stop (d) None of the above
    working

3. Network operating system runs on

(a) server   (b) every system in the network

(c) both (a) and (b)   (d) None of the above

4. Which technique is based on compile-time program transformation for accessing remote data in a distributed-memory parallel system?

(a) cache coherence scheme   (b) computation migration

(c) remote procedure call   (d) message passing

5. Logical extension of computation migration is

(a) process migration   (b) system migration

(c) thread migration   (d) data migration

6. Processes on the remote systems are identified by

(a) host ID   (b) host name and identifier

(c) identifier   (d) process ID

7. Which routing technique is used in distributed system?

(a) fixed routing   (b) virtual routing

(c) dynamic routing   (d) all of the mentioned

8. In distributed systems, link and site failure is detected by _____

(a) polling   (b) handshaking

(c) token passing   (d) none of the mentioned

9. The capability of a system to adapt the increased service load is called
_____

(a) scalability   (b) tolerance

(c) capacity   (d) None of the above

10. Internet provides _____ for remote login.

(a) telnet   (b) http

(c) ftp   (d) RPC

11. _____ of the distributed file system are dispersed among various machines
of distributed system.

(a) Clients   (b) Servers

(c) Storage devices   (d) all of the mentioned

12. Which one of the following hides the location where in the network the
file is stored?

(a) transparent distributed file system   (b) hidden distributed file system

(c) escaped distribution file system   (d) spy distributed file system

13. What are characteristic of Network Operating Systems ?

(a) Users are aware of multiplicity of   (b) They are transparent
machines

(c) They are simple to use   (d) All of the mentioned

14. How are access to resources of various machines is done ?

(a) Remote logging using ssh or   (b) Zone are configured for auto-
telnet   matic access

(c) FTP is not used   (d) All of the mentioned

15. What are characteristics of Distributed Operating system ?

(a) Users are aware of multiplicity of   (b) Access is done like local resources
machines

(c) Users are aware of multiplicity of   (d) They have multiple zones to ac-
machines   cess files

Objective type questions -answer

1:c 2:a 3:a 4:b 5:a 6:b 7:d 8:b 9:a 10:a 11:d 12:a 13:a 14:a 15:a

Review questions

1. Consider the parallel algorithm for solving the 0/1 knapsack problem. Derive the speedup and efficiency for this algorithm.

2. With regard to parallel computing, mention methods for containing interaction overheads.

3. Mention some of the parallel algorithm models.

4. Explain Asymptotic notations for parallel programs.

5. Explain programming using Message-passing Paradigm.

6. Explain programming using shared address space platforms.

7. Consider an SMP with a distributed shared-address-space. Consider a simple cost model in which it takes 10 ns to access local cache, 100 ns to access local memory, and 440 ns to access remote memory. A parallel program is running on this machine. The program is perfectly load balanced with 80% of all accesses going to local cache, 10% to local memory, and 10% to remote memory. What is the effective memory access time for this computation?

8. Write mutual exclusion algorithm for distributed computing.

9. Explain deadlock detection in distributed systems.

10. Explain authentication mechanisms in distributed systems.

11. For the multicast algorithm based on propagation trees, what is (i) tight upper bound on the number of multicast groups, and (ii) tight upper bound on the number of metagroups of the multicast groups ?

12. Design an experiment to determine the memory bandwidth of your computer and to estimate the caches at various levels of hierarchy. Justify your answer.

Critical thinking questions

1. Several desirable properties of a large-scale distributed system includes transparency of access, location, concurrency, replication, failure, migration,

performance and scaling. Analyze how each one of these properties applies to AWS.

2. What is in your opinion the critical step in the development of a systematic approach to all-or-nothing atomicity? What does a systematic approach means? What are the advantages of a systematic versus an ad-hoc approach to atomicity?

3. The support for atomicity affects the complexity of a system. Explain how the support for atomicity requires new functions/mechanisms and how these new functions increase the system complexity. At the same time, atomicity could simplify the description of a system; discuss how it accomplishes this.

4. The support for atomicity is critical for system features which lead to increased performance and functionality such as: virtual memory, processor virtualization, system calls, and user-provided exception handlers. Analyze how atomicity is used in each one of these cases.

5. Explain briefly how the publish-subscribe paradigm works and discuss its application to services such as bulletin boards, mailing lists, and so on. Outline the design of an event service based on this paradigm. Can you identify a Cloud service that emulates an event service?

6. Tuple spaces can be thought of as an implementation of a distributed shared-memory. Tuple-spaces have been developed for many programming languages including Java, Lisp, Python, Prolog, Smalltalk, and Tcl. Explain briefly how tuple spaces work. How secure and scalable are the tuple spaces you are familiar with, e.g., JavaSpaces?

Bibliography

[1] R. Chandra, L. Dagum, "Parallel Programming in OpenMP", Morgan Kaufmann Publishers, 2017.

[2] B. Codenotti, M. Leonicini, "Introduction to Parallel Processing", Addision-Wesley, 2016.

[3] M. Cole, "Structured Management of Parallel Computations", MIT press, 1990.

[4] I. Stoica, R. Morris, D. Karger, "Chord: A Scalable peer to peer lookup service for Internet Applications", IEEE Transactions on Networking, Vol. 11, No. 1, 2016, pp. 17-31.

[5] M. Guptha, M. H. Ammar, "Trade-offs between reliability and overheads in peer-to-peer reputation tracking", Computer Networks, Vol. 50, No. 4, 2018, pp. 501-522.

[6] B. Awerbuch, "Complexity of Network Synchronization", Journal of ACM, Vol. 34, No. 4, 2015, pp. 804-823.

[7] L. Lamport, "Concurrent reading and writing", Communications of ACM, Vol. 20, No. 11, 2017, pp. 806-811.

[8] "Reservoir project", *Available online at http://www.reservoir-fp7.eu,* Accessed on 09.05.2017

[9] Vouk M. A, "Issues, Research and Implementations in Information Technology Interfaces", *Proceedings of 30th IEEE International Conference on ITI*, Beijing, China, 2013, pp. 17-26.

[10] Kehey. K and Tsugua, "Sky computing", *Proceedings of IEEE International Conference on Internet Computing*, Singapore, 2015, pp. 11-16.

[11] Nurmi. D and Wolski. R, "The Eucalyptus Open-source Cloud-Computing System", *Proceedings of IEEE International Conference Cloud Computing and Its Applications*, Turin, Italy, 2014, pp. 234-236.

[12] C. Baun and M. Kunze, "Building a Private Cloud with Eucalyptus", *Proceedings of 11th IEEE International Conference on E-Science Workshops*, Munich, Germany, 2015, pp. 33-38.

[13] "Nimbus project", *Available online at http://www.nimbusproject.org*, Accessed on 09.06.2018.

[14] D. Nurmi, R. Wolski, C. Grzegorczyk, G. Obertelli, S. Soman, L. Youseff, and D. Zagorodnov, "The Eucalyptus Open-source Cloud-Computing System in Cloud Computing and Applications", *Proceedings of IEEE International Conference on Computability and Complexity in Analysis*, Tokyo, Japan, 2015, pp. 10-12.

[15] Harmer T and Wright P, "Provider-Independent Use of the Cloud", *Proceedings of IEEE International Workshop on Cloud Computing*, Vol. 5, No. 4, 2009, pp. 296-305.

[16] A. Sotomayor, I. Llorente, and A. Foster, "Virtual Infrastructure Management in Private and Hybrid Clouds", *Proceedings of IEEE International Workshop on Internet Computing*, Vienna, Austria, 2010, pp. 14-22.

[17] A. Sriram, "Simulation Tool Exploring Cloud-scale Data Centres", *Proceedings of IEEE International Conference on Cloud Computing*, Beijing, China, 2010, pp. 381-392.

3

Multicore Architectures

Learning Objectives

After reading this chapter, you will be able to

- Identify the significance of Multicore Architectures

- Compare and contrast the symmetric and asymmetric multiprocessors

- Analyze the parallelism in software and hardware

Multicore processor is viewed as the future production of microprocessor design. It is not only a solution for CPU speeds, but it also decreases the power consumption because many cores in a lower frequency collectively produce less heat dissipation than one core with their total frequency. From this point of view, Cloud computing is mostly built on top of multicore technologies. However, to fully take advantage of the computational capability and the other advantages of multicores, a lot of new techniques must be proposed and considered.

The significance of multicore processor architectures is shown in Fig. 3.1. As shown, multicore architecture has a computing component with two or more independent actual processing units (called "cores"), which are the units that read and execute program instructions. The instructions are ordinary CPU instructions (such as add, move data and branch), but the multiple cores can run multiple instructions at the same time, increasing overall speed for programs amenable to parallel computing. This chapter signifies parallel computing in the ongoing multicore era.

Preliminaries

The following are some of the terms and terminologies defined here for easier understanding in the remaining part of this chapter.

Hyperthreading : It is a technology used by microprocessors that allows a single microprocessor to act like two separate processors to the operating system and the application programs that use it.

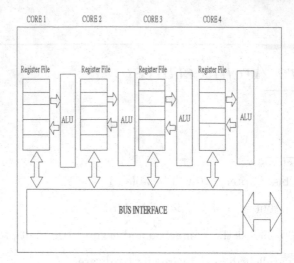

FIGURE 3.1
Multicore architecture

Multithreading : It is the ability of a program or an operating system process to manage its use by more than one user at a time and to even manage multiple requests by the same user without having multiple copies of the programs running in the computer.

L1, L2, L3 Cache : In most modern CPUs, the L1 and L2 caches are present on the CPU cores themselves, with each core getting its own cache. L2 cache holds data that is likely to be accessed by the CPU next. L3 (Level 3) cache is the largest cache memory unit, and also the slowest one. It can range between 4 MB to upwards of 50 MB.

Chip-multithreading : It is the capability of a processor to process multiple software threads and support simultaneous hardware threads of execution.

Pipelined processor : A scalar processor is one that acts on a single data stream. A pipelined (superscalar) CPU can execute more than one instruction per clock cycle. Because processing speeds are measured in clock cycles per second, a superscalar processor will be faster than a scalar processor rated at the same clock frequency given in MHz.

Concurrency : In computer science, concurrency refers to the ability of different parts or units of a program, algorithm, or problem to be executed out-of-order or in partial order, without affecting the final outcome.

Parallelism : Parallelism is basically a type of computation in which many computations or operations are carried out in parallel. This is done to achieve the speedup in computation.

3.1 Multicores in Cloud Computing

A multicore CPU combines multiple independent execution units into one processor chip in order to execute multiple instructions in a parallel fashion. The cores of a multicore processor are sometimes also denoted as processing element or computational engine. According to Flynn's taxonomy, the resulting systems are true multiple-instruction, multiple-data (MIMD) machines able to process multiple threads of execution at the same time.

Figure 3.2 shows multicore processor. In a multiprocessor system or a multicore processor (IntelQuad Core, Core two Duo etc.), each CPU core/processor has its own cache memory (data and program cache). Obviously, it varies by the exact chip model, but the most common design is for each CPU core to have its own private L1 data and instruction caches. On old and/or low-power CPUs, the next level of cache is typically a L2 unified cache that may be typically shared among all cores. Modern mainstream Intel CPUs (since the first-gen i7 CPUs, Nehalem) use 3 levels of cache. They are: (i) 32 kiB (kiloBytes) split L1i/L1d (i indicates instructions, d indicates data): private per-core, (ii) 256 kiB unified L2: private per-core, and (iii) Large unified L3: shared among all cores.

Last-level cache is a large shared L3. It is physically distributed between cores, with a slice of L3 going with each core on the ring bus that connects the core. Typically 1.5 to 2.25 MB of L3 cache is easily available, so a many-core Xeon might have a 36 MB L3 cache shared between all its cores. This is why a dual-core chip has 2 to 4 MB of L3, while a quad-core has 6 to 8 MB. Table 3.1 shows the difference between Single Core, Dual Core and Quad Core.

Multicore systems can provide high energy efficiency, since they can allow the clock frequency and supply voltage to be reduced together. This dramatically reduces power dissipation during periods, when full rate computation is not needed. As a simple example, assume one uniprocessor is capable of computing one application with clock rate F, and voltage V, and consuming power P. If using a dual core system and assuming the application can be partitioned into two cores without any overhead, each core only needs a clock rate $F/2$ and the voltage can be reduced accordingly; assuming a linear

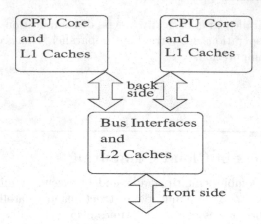

FIGURE 3.2
Multicore processor

relation between voltage and clock frequency and the voltage being reduced to $V/2$, the power dissipation of the dual core system will only be about $P/4$.

TABLE 3.1
Differences between Single Core, Dual Core and Quad Core

Single vs Multi Core		
Single Core	**Dual Core**	**Quad Core**
Features: CPU contains only one core to process different operations	Dual core CPU is a processor with two cores in a single IC	It contains two dual cores processors in a single IC
Applications: Word processing, checking mails, surfing the Internet, watching videos etc.	Flash enabled browsing, audio-video conference, chatting etc.	Voice-GPS systems, multi-player gaming, video editing etc.

There are two kinds of multicore processors: (i) Symmetric, and (ii) Asymmetric. A symmetric multicore processor is a processor which has multiple cores that are exactly the same. Examples are: Intel Core 2 and the Parallax Propeller. Every single core processor has the same architecture and the same capabilities. Each core has the same capabilities, so it requires that there is an arbitration unit to give each core a specific task. Software that uses techniques like multithreading makes the best use of a multicore processor like the Intel Core 2.

In an asymmetric multicore processor, as in Figure 3.3, the chip has multiple cores on-board. Each core has different capabilities. All the common components of the processor with vector/SIMD extensions (instruction control

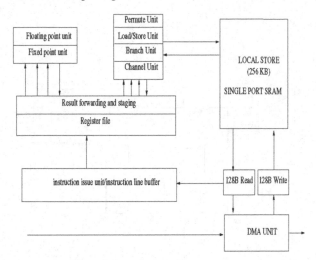

FIGURE 3.3
IBM Cell processor

unit, load and store unit, fixed-point integer unit, floating-point unit, vector unit, branch unit, virtual memory management unit) are hyper-threaded and support simultaneous threads. In addition to several registers, it also contains a DMA controller.

3.1.1 Parallel Hardware

A hardware-controlled parallel usage of execution unit components enable the execution of multiple processor instructions during one clock cycle. This approach of instruction-level parallelism (ILP) is limited, but widely realized approach in modern processor designs. A set of execution units can be put together to form a chip multi-processing (CMP) architecture. Most such processors contain separate L1 caches for each of the units, and a shared L2 cache for the whole processor chip. CMP forms the latest trend in processor design, even though earlier attempts already gained some experience with this approach. The different ways of exploiting parallelism inside one processor chip are then summarized as chip multi-threading (CMT) capabilities.

The next higher level is symmetric multiprocessing (SMP) with multiple processors, and ultimately the realization of a computer cluster as one multicomputer (Figure 3.4). Symmetric multiprocessors include two or more identical processors sharing a single main memory. The multiple processors may be separate chips or multiple cores on the same chip. Multiprocessors

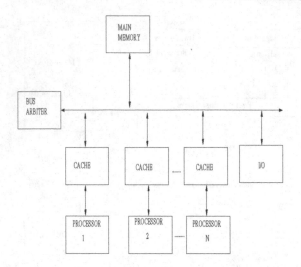

FIGURE 3.4
Symmetric multiprocessor systems

can be used to run more threads simultaneously or to run a particular thread faster. Running more threads simultaneously is easy; the threads are simply divided up among the processors. A bus arbiter is used in a multi-master bus system to decide which bus master will be allowed to control the bus for each bus cycle.

3.1.2   Parallel Software

Multicore frameworks are genuine parallel computers. The new pattern in this way, prompts the more extensive acknowledgment of somewhat notable and in part to a great extent programming issues. Industry and examination as of now concur upon the way that the current programming ideal models and dialects just as configuration designs do not adjust to present day processor design. This prompts the way that the scalability issue is going to be replaced by parallelisation issue.

Challenges that are well known to the parallel computing community are now the problem for every software developer. This is nothing less than a paradigm shift in education, training, and daily practice of software development. Developers must get a basic understanding of parallel programming from the very beginning, treating a sequential program only as special case. This relates not only to tools and languages, but also to design patterns, algorithmic thinking, testing strategies and software architecture principles.

3.2 Classes of Parallelism

Parallelism can be defined as the concept of dividing big problems into smaller ones, and solving the smaller problems by multiple processors at the same time. Parallelism is sometimes confused with concurrency even though it is not necessarily that every parallel processing is considered concurrent. Some types of parallelism like bit-level parallelism is not concurrent. Also, concurrency may involve working on different tasks like time-sharing multitasking, while parallelism means working on specific task in parallel computing.

Modern computer architecture implementation requires special hardware and software support for parallelism. Types of parallelism are Hardware parallelism and Software parallelism. Let us discuss about each of these.

Parallelism in Hardware: This refers to the type of parallelism defined by the machine architecture and hardware multiplicity. Hardware parallelism is a function of cost and performance tradeoffs. It displays the resource utilization patterns of simultaneously executable operations. It can also indicate the peak performance of the processors. One way to characterize the parallelism in a processor is by the number of instruction issues per machine cycle. In a modern processor, two or more instructions can be issued per machine cycle, e.g., i960CA is a 3-issue processor, capable of issuing three instruction per clock cycle.

Parallelism in a Uniprocessor: A uniprocessor system is defined as a computer system that has a single central processing unit that is used to execute computer tasks. As more and more modern software is able to make use of multiprocessing architectures, the term uniprocessor is used to distinguish the class of computers where all processing tasks share a single CPU. Most desktop computers are now shipped with multiprocessing architectures. As such, this kind of system uses a type of architecture that is based on a single computing unit. All operations (additions, multiplications, etc.) are thus done sequentially on the unit. Pipelining, Superscalar, Very Large Instruction Word (VLIW) etc., are examples.

SIMD instructions, Vector processors, GPUs: The 128-bit Vector/SIMD Multimedia Extension unit (VXU) operates concurrently with the fixed-point integer unit (FXU) and floating-point execution unit (FPU). Like PowerPC instructions, the Vector/SIMD Multimedia Extension instructions are 4 bytes long and word-aligned. The Vector/SIMD Multimedia Extension instructions support simultaneous execution on multiple elements that make up the 128-bit vector operands. Figure 3.5 shows concurrent execution of integer, floating-point and vector units to handle execution of several applications at the same time.

Instructions

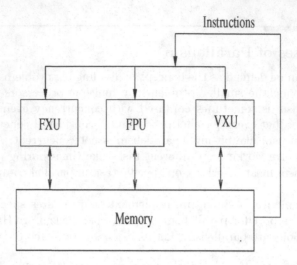

FIGURE 3.5
Concurrent execution of integer, floating-point and vector units

Multicomputers : It is a computer made up of several sub-computers. The term generally refers to an architecture in which each processor has its own memory rather than multiple processors with a shared memory. A multicore computer, although it sounds similar, would not be a multicomputer because the multiple cores share a common memory.

Parallelism in Software: It is defined by the control and data dependence of programs. The degree of parallelism is revealed in the program profile or in the program flow graph. Software parallelism is a function of algorithm, programming style, and compiler optimization. Different levels of parallelism are covered in Chapter 2.

3.3   Limitations of Multicore Architectures

A multicore processor consists of several cores which can execute different tasks independently. Due to the budget and chip area limit, the last level cache is usually shared among cores. Some of the limitations of multicore architectures are discussed as under.

- The Power Wall: Limit on the scaling of clock speeds, ability to handle on-chip heat has reached a physical limit.

- The Memory Wall: Need for bigger cache sizes, memory access latency still not in line with processor speeds.

- The Instruction Level Parallelism (ILP) Wall: Superlinear increase in complexity without linear increase in application performance.

- Imperfect scaling: Performance is dependent on serial code.

- Difficulty in software optimization: Easier to add cores, difficult for software to take advantage of them

- Maintaining concurrency over a number of cores.

To analyse the future microprocessor architecture, we must first understand the trade-offs among area (A), time (T), power (P) and technology (S). For a fixed process technology, higher performance (smaller T) inevitably requires more power (P) and area (A). The driving force in design innovation is the rapid advances in technology. As technology advances and feature size shrinks, the three design considerations benefit from one process generation to another, resulting in higher speed, smaller area and reduced power consumption.

Increasing the processor clock frequencies and using wide issue processor architectures have worked well to improve performance, but, recently have become significantly more challenging. Deeper pipelining is one of the key techniques to increase the clock frequency and performance, but the benefit of the deeper pipeline is eventually diminished when the inserted Flip-Flop's delay is comparable to the combinational logic delay. Moreover, deeper pipeline stage increases cycles-per-instruction (CPI) and impacts negatively to the system performance.

For most applications, power is the real price of performance. Processor power dissipation is the sum of dynamic switching power and static $leakage power := switching power + leakage power$. Switching power is directly proportional to capacitance and activity in the application programs. The leakage power increases exponentially with threshold voltage creating a major problem.

Performance-oriented designs take advantage of the decreased capacitance that the process technology provides to improve switching time accordingly. Large switching current, I, significantly improve switching times. If it is small compared with supply voltage, the switching current is roughly proportional to the square of the supply voltage. Thus, the switching frequency is roughly proportional to the supply voltage, which implies that there is a cubic relationship between T and power consumption, P, at the same feature size. Based on this cubic rule, it requires roughly eight times more power to reduce processing time by half. The trade-off of cost, performance and power consumption is fundamental to any system design. Although there are numerous design possibilities in this area-time-power continuum, the application always drives the optimum design point. Most high performance technologies, such as increasing clock frequencies and increasing processor issues (which mean increasing number of circuits and increasing capacitance), result in higher power consumption.

The computer architects face three major performance bottlenecks or walls discussed as follows.

Memory wall : The access time to memory depends on wire delay that is relatively constant with scaling. As processor clock rates increases, so do cache miss times, creating a memory wall on performance. So far, significant processor hardware support (such as increased cache size and branch tables) and an increased emphasis on memory performance, has helped designers manage cache misses and branches, reducing access to memory requirements and alleviating the impacts of memory wall. Nevertheless, the memory wall is still a serious issue.

Frequency wall: The maximum pipeline segment size (the number of logic gates in one pipeline stage) determines clock frequency. Reductions in segment size beyond a certain point are difficult, creating a frequency wall. Here, the question is how to improve performance without reducing segment size.

Power wall: Higher frequency implies greater power density. As clock speed increases, the cost of removing the resulting heat limits a design and its effectiveness at some point. Although these walls arise from physical constraints and application program behaviors, new processor and memory paradigms need to be explored to improve memory access and access predictability. For example, area concurrency and not clock frequency can help improve performance.

Summary

Multicore processor architecture entails silicon design engineers placing two or more execution cores, or computational engines, within a single processor package. This chapter discussed how 40 years of parallel computing research need to be considered in the upcoming multicore era. The future research must be driven from two sides, customer and service provider with a better expression of hardware structures, and a domain-specific understanding of software parallelism. To achieve parallelism joint efforts between hardware designer and software programmer are needed which could further upgrade computer systems performance.

Keywords

Multiprocessors	Hyperthreading
Multicores	Symmetric
Parallelism	Asymmetric

Objective type questions

1. Multicore systems of computer system are well suited for _____

(a) Database   (b) Web servers

(c) Designing   (d) both a and b

2. Foreground and background work application allow program to prompt for next command before previous command is _____

(a) Running   (b) Terminate

(c) Complete   (d) Update

3. Settimer() call function in windows operating system is Unix's function called _____

(a) chmod()   (b) timer()

(c) alarm()   (d) close()

4. Communication among computers via fixed paths or via some network facility, such a system is known as _____

(a) Branch   (b) Batch

(c) Cluster   (d) Bus

5. Basic unit of computer storage that is mainly used is _____

(a) byte   (b) nibble

(c) bit   (d) word

6. Efficiency for turning 115,000-volt power into 208-volt power which servers can use is _____

(a) 80%   (b) 82%

(c) 84%   (d) 89%

7. One of feature keys for achieving PUE of 1.23 was putting measurement devices such as _____

(a) Voltage divider   (b) Current divider

(c) Current transformers   (d) Voltage transformers

8. No of uplink ports, that are being used per rack-switch can varies from a minimum of _____

(a) 2   (b) 3

(c) 4   (d) 6

9. In Google terminology, Array is also called as _____

(a) Grid   (b) Stack

(c) Clusters   (d) Queue

10. A 1000 MB data transfer between disks within server, takes time of _____

(a) 2 seconds   (b) 5 seconds

(c) 8 seconds   (d) 10 seconds

11. If no node having a copy of a cache block, this technique is known as

(a) Uniform memory access   (b) Cached

(c) Un-cached   (d) Commit

12. A processor that continuously tries to acquire locks, spinning around a loop till it reaches its success, is known as _____

(a) Spin locks   (b) Store locks

(c) Link locks   (d) Store operational

13. Straight-forward model used for memory consistency, is called _____

(a) Sequential consistency   (b) Random consistency

(c) Remote node   (d) Host node

14. One assigned operation for building synchronized operations, is called the _____

(a) Atom exchange   (b) Atomic exchange

(c) Remote node   (d) Both a and b

15. To update cached copies of data item; is alternative protocol which is known as _____

(a) Write update   (b) Write broadcast protocol

(c) Read protocol   (d) both a and b

Objective type questions -answer

1:d   2:c   3:c   4:c   5:c   6:d   7:c   8:a   9:c   10:b   11:c   12:a   13:a   14:b 15:d

Review questions

1. Is there any general protocol or algorithm to start a multicore processor ? If so, how it works ?

2. Differentiate between multithreading and multicore.

3. Does multithreading use more than one core, if needed ?

4. Is there any downside to multi-core architecture? Is it possible that my application run slower ?

5. What dictates how many threads your computer has ?

6. What is the difference between multi-core architecture and Hyper-Threading (HT Technology) Technology ?

7. What is the impact of multi-core architecture on licensing? Is it going to increase software licensing costs ? If so, for which application, operating systems (OSes) or databases ?

8. What applications are good candidates to be moved from serial to multi-threaded to experience performance gains on multi-core systems?

9. Is any operating system (OS) particularly adept at benefiting from multi-core architecture ? Which one(s) and why?

10. If I do nothing, will I benefit from multi-core architecture? How? Or will my application's performance suffer ?

Critical thinking questions

1. Compare the Oracle Cloud offerings (see https://cloud.oracle.com) with the cloud services provided by Amazon, Google, and Microsoft.
2. What Intel Software Development Products can help to create threaded software?

3. How much work/time/effort will it take me to thread my application? How can I determine if I should thread my application (or not) ?

4. Besides using threaded applications, when else might end users experience a performance gain on multi-core systems?

Bibliography

[1] E. Caron, F. Desprez, and A. Muresan, "Pattern Matching Based Forecasting of Non-Periodic Repetitive Behaviour for Cloud Clients", *Journal of Grid Computing*, Vol. 9, No. 5, 2011, pp. 49-64.

[2] Y. Shi, X. Jiang, and K. Ye, "An Energy-Efficient Scheme for Cloud Resource Provisioning Based on CloudSim", *Proceedings of IEEE International Conference on Cluster Computing*, Taiwan, Japan, 2011, pp. 595-599.

[3] P. Dinda, "Online Prediction of the Running Time of Tasks", *Proceedings of 10th IEEE International Symposium on High Performance Distributed Computing*, Washington, USA, 2011, pp. 383-394.

[4] W. Smith, I. Foster, and V. Taylor, "Predicting Application Run Times Using Historical Information", *Elsevier Journal of Parallel and Distributed Computing*, Vol. 64, No. 9, 2010, pp. 1007-1016.

[5] C. Lin, C. Lu, Y. Chen, J. Li, "Resource Allocation in Cloud Virtual Machines Based on Empirical Service Traces", *Journal of Communication Systems*, Vol. 3, No. 5, 2013, pp. 98-102.

[6] J. Huang and Li. C, "Resource Prediction Based on Double Exponential Smoothing in Cloud Computing", *Proceedings of IEEE International Conference on Consumer Electronics, Communications and Networks*, Shanghai, China, 2012, pp. 2056-2060.

[7] R. Hu, J. Jiang, G. Liu, and L. Wang, "CPU Load Prediction Using Support Vector Regression and Kalman Smoother for Cloud", *Proceedings of 33rd IEEE International Conference on Distributed Computing Systems*, Ohio, USA, 2013, pp. 88-92.

[8] D. Williams, H. Jamjoom, and Y. Liu, "Overdriver: Handling Memory Overload in an Oversubscribed Cloud", *Proceedings of 7th ACM SIGPLAN/SIGOPS International Conference on Virtual Execution Environments*, CA, USA, 2013, pp. 205-216.

[9] T. Wood, G. Tarasuk Levin, and P. Shenoy, "Memory buddies: Exploiting Page Sharing for Smart Colocation in Virtualized Data Centers", *Proceedings of the ACM SIGPLAN/SIGOPS International Conference on Virtual Execution Environments*, CA, USA, 2012, pp. 31-40.

61

[10] E. Tasulos and K. Begnnum, "Bayllocator: A Proactive System to Predict Server Utilization and Dynamically Allocate Memory Resources Using Bayesian Networks and Ballooning", *Proceedings of 26th Large Installation System Administration Conference*, California, USA, 2012, pp. 111-124.

[11] Sheng Di, Derrick Kondoa, and Walfredo Cirne, "Google Hostload Prediction Based on Bayesian Model with Optimized Feature Combination", Élsevier Journal of Parallel and Distributed Computing, Vol. 74, No. 1, 2014, pp. 1820-1832.

[12] C. Viswanath and C. Valliyammai, "CPU Load Prediction Using ANFIS for Grid Computing", *Proceedings of IEEE International Conference on Advances in Engineering, Science and Management*, Bengaluru, India, 2012, pp. 343-348.

[13] M. Dashevskiy and Z. Luo, "Time Series Prediction with Performance Guarantee", *Journal of IET Communications*, Vol. 5, No. 8, 2010, pp. 1041-1051.

[14] M. Sarkar, "Resource Requirement Prediction Using Clone Detection Technique", *Journal of Future Generation Computer Systems*, Vol. 29, No. 4, 2013, pp. 936-952.

[15] J. Kupferman, "Scaling Into the Cloud", Available Online at http://www.cs.ucsb.edu/jkupferman/docs/ScalingIntoTheClouds.pdf, Accessed on 26.06.2014.

[16] Miguel A. Lopez-Carmona, Ivan Marsa-Maestre, and Mark Klein, "Consensus Policy Based Multi-Agent Negotiation", *Proceedings of 14th IEEE International Conference on PRIMA*, Wollongong, Australia, 2011, pp. 159-173.

[17] Mikoto Okumura and Katsuhide Fujita, "Implementation of Collective Collaboration Support System based on Automated Multi-Agent Negotiation", *Text Book on Complex Automated Negotiations: Theories, Models, and Software Competitions* , Springer, Vol. 43, 2013, pp. 125-141.

[18] Ivan Marsa-Maestre, Miguel A. Lopez-Carmona, and Mark Klein, "A Scenario Generation Framework for Consistent Comparison of Negotiation Approaches", *Proceedings of 14th IEEE International Conference on PRIMA*, Wollongong, Australia, 2011, pp. 174-183.

[19] Bo Antony, Victor Lesser, David Irwin, and Michael Zink, "Automated Negotiation with Decommitment for Dynamic Resource Allocation in Cloud Computing", *Proceedings of 9th IEEE International Conference on Autonomous Agents and Multiagent Systems*, Beijing, China, 2013, pp. 19-27.

[20] Moustapha Tahir Ateib, "Agent Based Negotiation In E-commerce", *Journal of Information Technology*, Vol. 2, No. 1, 2010, pp. 861-868.

4

Virtualization

After reading this chapter, you will be able to

- Summarize the significance of Virtualization

- Appraise the Virtualization techniques

- Realize the pros and cons of Virtualization

The term "virtualization" was coined in the 1960s to refer to a virtual machine (sometimes called "pseudo machine"), a term which itself dates from the experimental IBM M44/44X system. The creation and management of virtual machines (VM) has been called "platform virtualization", or "server virtualization", more recently. Platform virtualization is performed on a given hardware platform by host software (a control program), which creates a simulated computer environment, a VM for its guest software. The guest software is not limited to user applications; many hosts allow the execution of complete operating systems. The guest software executes as if it were running directly on the physical hardware, with several notable caveats.

In this chapter we discuss about the virtulization technology, different techniques of virtualization, pros and cons of virtulaization.

Preliminaries

The following are some of the terms and terminologies defined here for easier understanding in the remaining part of this chapter.

Virtualization : Virtualization is a broad term that refers to the abstraction of computer resources. Virtualization hides the physical characteristics of computing resources from their users, be thee applications, or end users.

Hypervisor : A hypervisor, also known as a virtual machine monitor or VMM, is software that creates and runs virtual machines (VMs). A hypervisor allows one host computer to support multiple guest VMs by virtually sharing its resources, such as memory and processing.

Emulation : Emulation is the use of an application program or device to

65

imitate the behavior of another program or device.

Binary translation : Binary translation is a form of binary recompilation where sequences of instructions are translated from a source instruction set to the target instruction set.

Paravirtualization : Para-virtualization is a virtualization technique that presents a software interface to the virtual machines which is similar, yet not identical to the underlying hardware-software interface.

Software virtualization : It is just like a virtualization but able to abstract the software installation procedure and create virtual software installations. Virtualized software is an application that will be "installed" into its own self-contained unit.

Hardware virtualization : Hardware virtualization enables multiple copies of the same or different operating systems to run in the computer and prevents the OS and its applications in one VM from interfering with the OS and applications in another VM.

Network and Storage Virtualization: In a network, virtualization consolidates multiple devices into a logical view so they can be managed from a single console. Virtualization also enables multiple storage devices to be accessed the same way no matter their type or location.

Containers : Containers are the products of operating system virtualization. They provide a lightweight virtual environment that groups and isolates a set of processes and resources such as memory, CPU, disk, etc., from the host and any other containers.

OS Virtualization : Under the control of one operating system, a server is split into containers where each one handles an application.

Fig. 4.1 shows the virtualization architecture. It shows the guest operating system which has the capability to invoke VM instances. The VM is a software computer that, like a physical computer, runs an operating system and applications. The hypervisor serves as a platform for running virtual machines and allows for the consolidation of computing resources. Both these layers are supported by the host operating system layer.

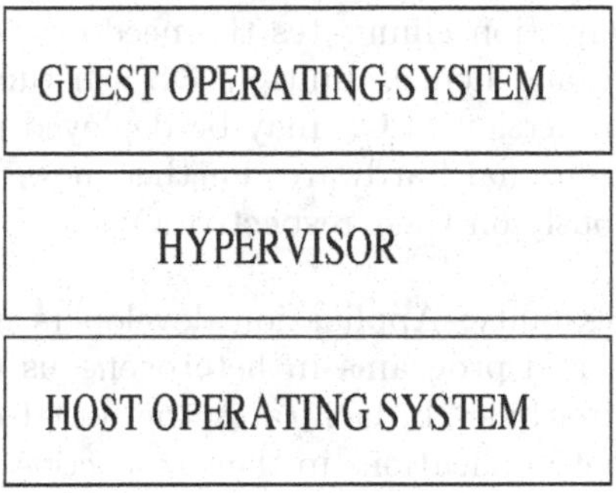

GUEST OPERATING SYSTEM

HYPERVISOR

HOST OPERATING SYSTEM

FIGURE 4.1
Virtualization architecture

4.1 Virtualization Technology

Any discussion of Cloud computing typically begins with virtualization. Virtualization is means of using computer resources to imitate other computer resources or whole computers. It separates resources and services from the underlying physical delivery environments.

Characteristics of Virtualization in Cloud Computing

Virtualization has several characteristics that make it ideal for Cloud computing, which are discussed as follows:

Partitioning : In virtualization, many applications and operating systems (OS) are supported in a single physical system by partitioning (separating) the available resources.

Isolation : Each VM is isolated from its host physical system and other virtualized machines. Because of this isolation, if one virtual-instance crashes, it does not affect the other VMs. In addition, data is not shared between one virtual container and another.

Encapsulation : A VM can be represented (and even stored) as a single file, so its identification is easy, based on the services that it provides. In essence, the encapsulated process could be a business service. This encapsulated VM can be presented to an application as a complete entity. Therefore, encapsulation can protect each application in order to stop its interference with another application.

Consolidation : Virtualization eliminates the need of a dedicated single system to one application and hence, multiple OS can run in the same server. Both old and advanced version of OS may be deployed in the same platform without purchasing additional hardware. Further, new required applications may be run simultaneously on their respective OS.

Easier development flexibility: Application developers may able to run and test their applications and programs in heterogeneous OS environments on the same virtualized machine. It facilitates the VM to host heterogeneous OS. Isolation of different applications in their respective virtual partition also helps the developers.

Migration and cloning: VM can be moved from one site to another to balance the workload. As the result of migration, users can access updated hardware as well as make recovery from hardware failure. Cloned VMs are easy to deploy in the local sites as well as remote sites.

Stability and security: In a virtualized environment, host OS hosts different types of multiple guest OS containing multiple applications. Each VM is isolated from each other and they do not interfere into the other's work, which in turn helps the security and stability aspect.

4.2   Virtualization Platforms

Platform virtualization software, specifically emulators and hypervisors, are software packages that emulate the whole physical computer machine, often providing multiple virtual machines on one physical platform. Here, we discuss the basic information about platform virtualization hypervisors, namely Xen and VMware hypervisors.

4.2.1   Xen Virtualization

It is available for the Linux kernel, and is designed to consolidate multiple OS to run on a single server, normalize hardware accessed by the OS, isolate misbehaving applications, and migrate running OS instances from one physical server to another. Recent advances in virtualization technologies, such as enabling data centers to consolidate servers, normalize hardware resources and isolate applications on the same physical server, are driving rapid adoption of server virtualization in Linux environments.

Fig. 4.2 shows Xen hypervisor architecture. The Xen hypervisor is the basic abstraction layer of software that sits directly on the hardware below any

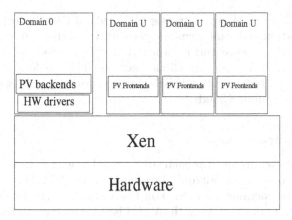

FIGURE 4.2
Xen Hypervisor architecture

operating systems. It is responsible for CPU scheduling and memory partitioning of the various virtual machines running on the hardware device. Domain 0 , a modified Linux kernel, is a unique virtual machine running on the Xen hypervisor that has special rights to access physical I/O resources as well as interact with the other virtual machines (Domain U : Para Virtual (PV) and Hardware Virtual Machine (HVM) Guests) running on the system. All Xen virtualization environments require Domain 0 to be running before any other virtual machines can be started.

All paravirtualized virtual machines running on a Xen hypervisor are referred to as Domain U PV Guests and run modified Linux operating systems, Solaris, FreeBSD and other UNIX operating systems. All fully virtualized machines running on a Xen hypervisor are referred to as Domain U HVM Guests and run on standard Windows or any other unchanged operating system.

Domain 0 is the initial domain started by the Xen hypervisor on boot. Domain 0 is a privileged domain that starts first and manages the Domain U unprivileged domains. The Xen hypervisor is not usable without Domain 0. This is essentially the "host" operating system (or a "service console"). As a result, Domain 0 runs the Xen management toolstack, and has special privileges, like being able to access the hardware directly.

Domain 0 has drivers for hardware, and it provides Xen virtual disks and network access for guests each referred to as a domU (unprivileged domains). For hardware that is made available to other domains, like network interfaces and disks, it will run the BackendDriver, which multiplexes and forward the hardware requests from the FrontendDriver in each Domain U. Unless Driver-Domain's are being used or the hardware is passed to the domain U, the domain 0 is responsible for running all of the device drivers for the hardware.

The Xen Cloud platform addresses the needs of Cloud providers by combining the isolation and multi-tenancy capabilities of the Xen hypervisor with enhanced security, storage and network virtualization technologies to offer a rich set of virtual infrastructure Cloud services. The platform also addresses user requirements for security, availability, performance and isolation across both private and public Clouds.

4.2.2   VMware

The traditional mainframe approach runs virtual machines in a less privileged mode in order to allow the Virtual Machine Monitor (VMM) to regain control on privileged instructions, and relies on the VMM to virtualize and interface directly to the I/O devices. Also, the VMM is in complete control of the entire machine. This approach does not apply as easily to PCs for the following reasons.

Non-virtualizable processor: The Intel IA-32 processor architecture is not naturally virtualizable. Because the IA-32 processor does not meet this condition, it is not possible to virtualize the processor by simply executing all virtual machine instructions in a less privileged mode.

PC hardware diversity: There is a large diversity of devices that may be found in PCs. This is a result of the PC's "open" architecture. In a traditional implementation, the virtual machine monitor would have to manage these devices. This would require a large programming effort to provide device drivers in the VMM for all supported PC devices.

Pre-existing PC software: Unlike mainframes that are configured and managed by experienced system administrators, desktop and workstation PC's are often pre-installed with a standard OS set up and managed by the end-user. In this environment, it is extremely important to allow a user to adopt virtual machine technology without losing the ability to continue using the existing OS and applications. It would be unacceptable to completely replace an existing OS with a virtual machine monitor.

VMware Workstation has a hosted architecture that allows it to co-exist with a pre-existing host operating system, and rely upon that operating system for device support. Figure 4.3 illustrates the components of this hosted architecture. VMware Workstation installs like a normal application on an operating system, known as the host operating system. When run, the application portion (VM APP) uses a driver loaded into the host operating system (VM DRIVER) to establish the privileged virtual machine monitor component (VMM) that runs directly on the hardware. From then on, a given physical processor is executing either the host world or the VMM world, with the VM Driver facilitating the transfer of control between the two worlds. A world

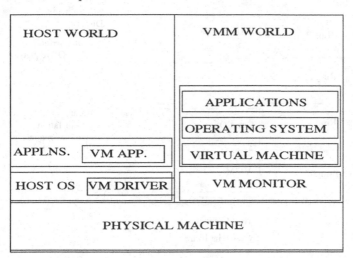

FIGURE 4.3
VMWare's hosted VM

switch between the VMM and the host worlds involves saving and restoring all user and system visible state on the CPU, and is thus more heavyweight than a normal process switch.

4.3 Virtualization Techniques

When deciding on the best approach to implementing virtualization, it is important to have a clear understanding of the different virtualization solutions which are currently available. We discuss the four virtualization techniques that are commonly used today, namely hypervisor, guest operating system, shared kernel and kernel level.

4.3.1 Hypervisor Virtualization

The x86 family of CPUs provide a range of protection levels, also known as rings in which code can execute (Figure 4.4). Ring 0 has the highest level privilege and it is in this ring that the OS kernel normally runs. Code executing in ring 0 is said to be running in system space, kernel mode or supervisor mode. All other code such as applications running on the OS operates in less privileged rings, typically ring 3. Under hypervisor virtualization, a program known as a hypervisor (also known as a type 1 virtual machine monitor or

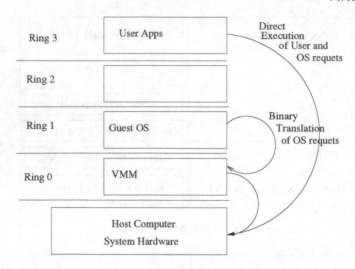

FIGURE 4.4
Binary translation approach to x86 virtualization

VMM) runs directly on the hardware of the host system in ring 0. The task of this hypervisor is to handle resource and memory allocation for the VM in addition to providing interfaces for higher level administration and monitoring tools.

Clearly, with the hypervisor occupying ring 0 of the CPU, the kernels for any guest operating systems running on the system must run in less privileged CPU rings. Unfortunately, most operating system kernels are written explicitly to run in ring 0 for the simple reason that they need to perform tasks that are only available in that ring, such as the ability to execute privileged CPU instructions and directly manipulate memory. A number of different methodologies to this problem have been devised in recent years, each of which is given here.

- Binary translation

- Full virtualization

- Paravirtualization

- Hardware assisted virtualization

Binary translation is one specific approach to implement full virtualization that does not require hardware virtualization features. It involves examining the executable code of the virtual guest for unsafe instructions, translating

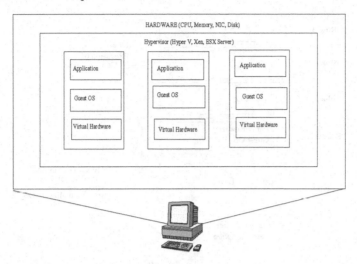

FIGURE 4.5
Logical Diagram of Full Virtualization

these into safe equivalents, and then executing the translated code. Alternatives to binary translation are binary patching, and full system emulation.

Full Virtualization provides support for unmodified guest OS. The term unmodified refers to OS kernels which have not been altered to run on a hypervisor and therefore still execute privileged operations as though running in ring 0 of the CPU. The hypervisor provides CPU emulation to handle and modify privileged and protected CPU operations made by unmodified guest OS kernels. Unfortunately, this emulation process requires both time and system resources to operate, resulting in inferior performance levels when compared to those provided by para-virtualization.

Full Virtualization Using Binary Translation: This approach relies on binary translation to trap (into the VMM) and to virtualize certain sensitive and non-virtualizable instructions with new sequences of instructions that have the intended effect on the virtual hardware. Meanwhile, user level code is directly executed on the processor for high performance virtualization, as shown in Figure 4.5.

Paravirtualization is a technique that presents a software interface to virtual machine that is similar, but not identical to that of the underlying hardware. The intent of the modified interface is to reduce the portion of the guest's execution time spent performing operations which are substantially more difficult to run in a virtual environment compared to a non-virtualized

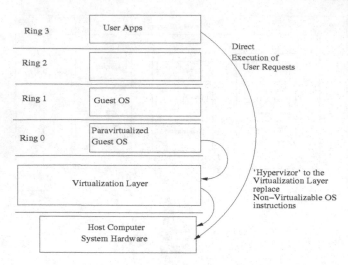

FIGURE 4.6
Paravirtualization approach to x86 virtualization

environment. The paravirtualization provides specially defined "hooks" to allow the guest(s) and host to request and acknowledge these tasks, which would otherwise be executed in the virtual domain (where execution performance is worse). A successful paravirtualized platform may allow the virtual machine monitor (VMM) to be simpler (by relocating execution of critical tasks from the virtual domain to the host domain), and/or reduce the overall performance degradation of machine-execution inside the virtual-guest.

Paravirtualization requires the guest operating system to be explicitly ported for the para-API. A conventional OS distribution that is not paravirtualization-aware cannot be run on top of a paravirtualizing VMM. However, even in cases where the operating system cannot be modified, components may be available that enable many of the significant performance advantages of paravirtualization.

Under paravirtualization, the kernel of the guest OS is modified specifically to run on the hypervisor as shown in Figure 4.6. This typically involves replacing any privileged operations that only runs in ring 0 of the CPU with calls to the hypervisor (known as hypercalls). The hypervisor in turn performs the task on behalf of the guest kernel. This typically limits support to open source OS such as Linux which is freely altered and proprietary OS where the owners agree to make the necessary code modifications to target a specific hypervisor. These issues not withstanding, the ability of the guest kernel to communicate directly with the hypervisor results in greater performance levels than other virtualization approaches.

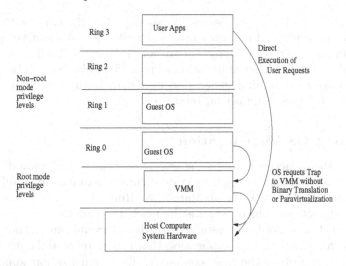

FIGURE 4.7
Hardware assisted approach to x86 virtualization

Hardware Virtualization

In computing, hardware-assisted virtualization is a plaform virtualization approach that enables efficient full virtualization using help from hardware capabilities, primarily from the host processors. Full virtualization is used to simulate a complete hardware environment, or virtual machine, in which an unmodified guest operating system (using the same instruction set as the host machine) executes in complete isolation. Hardware-assisted virtualization was added to x86 processors (Intel VT-x or AMD-V) in 2005 and 2006, respectively. Hardware-assisted virtualization is also known as accelerated virtualization; Xen calls it hardware virtual machine (HVM), Virtual Iron calls it native virtualization.

Hardware virtualization leverages virtualization features built into the latest generations of CPUs from both Intel and AMD. These technologies, known as Intel VT and AMD-V, respectively, provide extensions necessary to run unmodified guest VM without the overheads inherent in full virtualization CPU emulation. In very simplistic terms, these new processors provide an additional privilege mode above ring 0, in which the hypervisor operates essentially leaving ring 0 available for unmodified guest OS. Figure 4.7 illustrates hardware assisted approach to x86 virtualization.

Intel's Virtualization Technology (VT-x) (e.g. Intel Xeon) and AMD's AMD-V both target privileged instructions with a new CPU execution mode feature that allows the VMM to run in a new root mode below ring 0, also referred to as Ring 0P (for privileged root mode) while the Guest OS runs in Ring 0D (for de-privileged non-root mode). Privileged and sensitive calls are set to automatically trap to the hypervisor and handled by hardware, removing the need for either binary translation or para-virtualization. VMware only takes advantage of these first generation hardware features in limited cases such as for 64-bit guest support on Intel processors.

4.3.2   Guest OS Virtualization

Guest OS virtualization is perhaps the easiest concept to understand. In this scenario, the physical host computer system runs a standard unmodified OS such as Windows, Linux, Unix or MacOS X. Running on this OS is a virtualization application which executes in much the same way as any other application such as a word processor or spreadsheet would run on the system. It is within this virtualization application that one or more VMs are created to run the guest OS on the host computer. The virtualization application is responsible for starting, stopping and managing each VM and essentially controlling access to physical hardware resources on behalf of the individual VMs. The virtualization application also engages in a process known as binary rewriting which involves scanning the instruction stream of the executing guest system and replacing any privileged instructions with safe emulations. This has the effect of making the guest system think that, it is running directly on the system hardware, rather than in a VM inside an application.

Some examples of guest OS virtualization technologies include VMware Server and VirtualBox. Figure 4.8 provides an illustration of guest OS-based virtualization. Here, the guest OS operate in VMs within the virtualization application which, in turn, runs on top of the host OS in the same way as any other application. Clearly, the multiple layers of abstraction between the guest OS and the underlying host hardware are not conducive to high levels of VM performance. This technique does, however, have the advantage that no changes are necessary to either host or guest OS and no special CPU hardware virtualization support is required.

4.3.3   Shared Kernel Virtualization

The structure of shared kernel virtualization is illustrated in Figure 4.9. Shared kernel virtualization (also known as system level or operating system virtualization) takes advantage of the architectural design of Linux and UNIX based OS. In order to understand how shared kernel virtualization works, we need to first understand the two main components of Linux or UNIX OS. At the core of the OS is the kernel. The kernel, in simple terms, handles all the interactions between the OS and the physical hardware. The second key component is the root file system which contains all the libraries, files and utilities necessary for the OS to function. Under shared kernel virtualization, each of the virtual

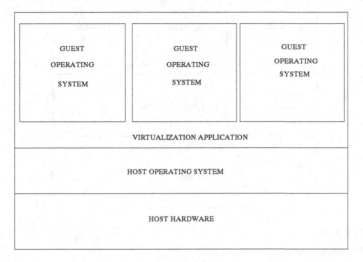

FIGURE 4.8
Guest OS virtualization

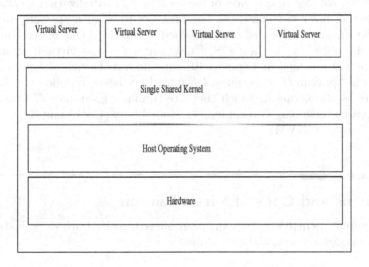

FIGURE 4.9
Shared kernel virtualization

guest systems (virtual server in Figure 4.9) have their own root file system, but share the kernel of the host OS.

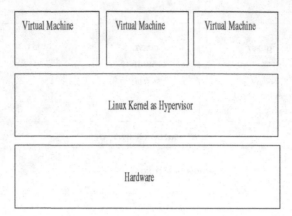

FIGURE 4.10
Kernel virtualization

4.3.4   Kernel Level Virtualization

Figure 4.10 provides an overview of the kernel level virtualization architecture. Under kernel level virtualization, the host OS runs on a specially modified kernel which contains extensions designed to manage and control multiple VMs each containing a guest OS. Unlike shared kernel virtualization, each guest (virtual machine in Figure 4.10) runs its own kernel, although similar restrictions apply in that the guest OS must have been compiled for the same hardware as the kernel in which they are running. Examples of kernel level virtualization technologies include User Mode Linux (UML) and Kernel-based Virtual Machine (KVM).

4.4   Pros and Cons of Virtualization

Most common benefits of virtualization offered up by both organizations as well as vendors are seen to be the following.

It is cheaper: Because virtualization does not require actual hardware components to be used or installed, IT infrastructures find it to be a cheaper system to implement. There is no longer a need to dedicate large areas of space and huge monetary investments to create an on-site resource. We just purchase the license or the access from a third-party provider and begin to work, just as if the hardware were installed locally.

It keeps costs predictable: Because third-party providers typically provide virtualization options, individuals and corporations can have predictable costs for their information technology needs.

It reduces the workload: Most virtualization providers automatically update their hardware and software that will be utilized. Instead of sending people to do these updates locally, they are installed by the third-party provider.

It offers a better uptime: Thanks to virtualization technologies, uptime has improved dramatically. Some providers offer an uptime that is 99.9999%. Even budget-friendly providers offer uptime at 99.99% today.

It allows for faster deployment of resources: Resource provisioning is fast and simple when virtualization is being used. There is no longer a need to set up physical machines, create local networks, or install other information technology components.

It promotes digital entrepreneurship: Before virtualization occurred on a large scale, digital entrepreneurship was virtually impossible for the average person. Thanks to the various platforms, servers, and storage devices that are available today, almost anyone can start their own side hustle or become a business owner.

It provides energy savings: For most individuals and corporations, virtualization is an energy-efficient system. Because there are not local hardware or software options being utilized, energy consumption rates can be lowered. Instead of paying for the cooling costs of a data center and the operational costs of equipment, funds can be used for other operational expenditures over time to improve virtualization's overall Return-on-Investment (ROI) .

Disadvantages of Virtualization There can be some downsides to virtualization, as well. They are discussed below.

It can have a high cost of implementation: The cost for the average individual or business when virtualization is being considered will be quite low. For the providers of a virtualization environment, however, the implementation costs can be quite high.

It creates a security risk: Information is our modern currency. If you have it, you can make money. If you do not have it, you may be ignored. Because data is crucial to the success of a business, it is targeted frequently.

It creates an availability issue: The primary concern that many have with virtualization is what will happen to their work should their assets not be available. If an organization cannot connect to their data for an extended period of time, they will struggle to compete in their industry.

It creates a scalability issue: Although you can grow a business or opportunity quickly because of virtualization, you may not be able to become as large as you would like. You may also be required to be larger than you want to be when first starting out.

It requires several links in a chain that must work together cohesively: If you have local equipment, then you are in full control of what you can do. With virtualization, you lose that control because several links must work together to perform the same task.

It takes time: Although you save time during the implementation phases of virtualization, it costs users time over the long-run when compared to local systems.

It still has limitations: Not every application or server is going to work within an environment of virtualization. That means an individual or corporation may require a hybrid system to function properly.

The advantages and disadvantages of virtualization show us that it can be a useful tool for individuals, small and medium businesses, entrepreneurs, and corporations when it is used properly. Because it is so easy to use, however, some administrators begin adding new servers or storage for everything and that creates sprawl. By staying disciplined and aware of communication issues, many of the disadvantages can be tempered, which is why this is such an effective modern system.

Summary

Nowadays, virtualization is a technology that is applied for sharing the capabilities of physical computers by splitting the resources among OSs. The concept of Virtual Machines (VMs) started back in 1964 with a IBM project called CP/CMS system. Currently, there are several virtualization techniques that can be used for supporting the execution of entire operating systems. In this chapter, we classified the virtualization techniques from the OS view. First, we discussed two techniques that executes modified guest OSs: operating system-level virtualization and para-virtualization. Second, we discussed two techniques that executes unmodified guest OSs: binary translation and hardware assisted. Finally, we highlighted on the prons and cons of virtualization.

Keywords

Virtualization	Binary translation
Paravirtualization	Software virtualization
Hardware virtualization	

Objective type questions

1. Point out the wrong statement :

(a) Abstraction enables the key benefit of Cloud computing: shared, ubiquitous access

(b) Virtualization assigns a logical name for a physical resource and then provides a pointer to that physical resource when a request is made

(c) All Cloud computing applications combine their resources into pools that can be assigned on demand to users

(d) Both statements in (a) & (b)

2. Which of the following type of virtualization is also characteristic of Cloud computing ?

(a) Storage

(b) Application

(c) CPU

(d) All of the above

3. The technology used to distribute service requests to resources is referred to as : ?

(a) Load performing

(b) Load scheduling

(c) Load balancing

(d) All of the above

4. Point out the correct statement :

(a) A client can request access to a Cloud service from any location

(b) A Cloud has multiple application instances and directs requests to an instance based on conditions

(c) Computers can be partitioned into a set of virtual machines with each machine being assigned a workload

(d) All of the above

5. Which of the following software can be used to implement load balancing ?

(a) Apache mod_balancer

(b) Apache mod_proxy_balancer

(c) F0's BigIP

(d) All of the above

6. Which of the following network resources can be load balanced ?

(a) Connections through intelligent switches

(b) DNS

(c) Storage resources

(d) All of the above

7. Which of the following is a more sophisticated load balancer ?

(a) Workload managers

(b) Workspace managers

(c) Rackserve managers

(d) All of the above

8. An _____ is a combination load balancer and application server that is a server placed between a firewall or router.

(a) . ABC

(b) ACD

(c) ADC

(d) All of the above

9. Point out the wrong statement

(a) Load balancing virtualizes systems and resources by mapping a logical address to a physical address

(b) Multiple instances of various Google applications are running on different hosts

(c) Google uses hardware virtualization

(d) All of the above

10. Which of the following is another name for system virtual machine ?

(a) Hardware virtual machine

(b) Software virtual machine

(c) Real machine

(d) None of the above

11. Which of the following provide system resource access to virtual machines ?

(a) VMM

(b) VMC

(c) VNM

(d) All of the above

12. Point out the correct statement :

(a) A virtual machine is a computer that is walled off from the physical computer that the virtual machine is running on

(b) Virtual machines provide the capability of running multiple machine instances, each with their own operating system

(c) The downside of virtual machine technologies is that having resources indirectly addressed means there is some level of overhead

(d) All of the above

13. An operating system running on a Type _____ VM is a full virtualization.

(a) 1   (b) 2

(c) 3   (d) All of the above

14. Which of the following is Type 1 Hypervisor ?

(a) Wind River Simics   (b) Virtual Server 2005 R2

(c) KVM   (d) LynxSecure

15. Which of the following is Type 2 VM ?

(a) VirtualLogix VLX   (b) VMware ESX

(c) Xen   (d) LynxSecure

Objective type questions -answer

1:c 2:d 3:c 4:d 5:b 6:d 7:a 8:c 9:c 10:a 11:a 12:d 13:a 14:d
15:c

Review questions

1. What is Virtualization ?

2. What are the types of hardware virtualization?

3. What are the benefits of virtualization?

4. What is the purpose of a Hypervisor?

5. How ESX server related to VMWare?

6. What is the difference between ESX and GSX server?

7. How VMWare Kernel different from other kernels?

8. What are the features provided by VMWare for easy access?

9. What are the different components used in VMWare infrastructure?

10. Compare and contrast different techniques of virtualization

Critical thinking questions

1. Identify the milestones in the evolution of operating systems during the half century from 1960 to 2010 and comment on the statement "VMMs give operating system developers another opportunity to develop functionality no longer practical in today's complex systems, where innovation moves at a geologic pace".

2. Virtualization simplifies the use of resources, isolates users from one another, supports replication and mobility, but exacts a price in terms of performance and cost. Analyze each one of these aspects for: (i) memory virtualization, (ii) processor virtualization, and (iii) virtualization of a communication channel.

3. Virtualization of the processor combined with virtual memory management pose multiple challenges; analyze the interaction of interrupt handling and paging.

4. In 2012 Intel and HP announced that Itanium architecture will be discontinued. Review the architecture discussed and identify several possible reasons for this decision.

Bibliography

[1] Endriss Ulle, Maudet Nicolas, Sadri Fariba, and Toni Francesca, "Negotiating Socially Optimal Allocations of Resources", *Journal of Artificial Intelligence*, Vol. 25, No. 5, 2006, pp. 315-348.

[2] R. Buyya, D. Abramson, J. Giddy, and H. Stockinger, "Economic Models for Resource Management and Scheduling in Grid Computing", *Journal of Concurrency and Computation: Practice and Experience*, Vol. 14, No. 13, 2012, pp. 1507-1542.

[3] F. Lang, "Developing Dynamic Strategies for Multi-issue Automated Contracting in the Agent Based Commercial Grid", *Proceedings of IEEE International Symposium on Cluster Computing and Grid*, Guangdong, China, 2013, pp. 342-349.

[4] H. Gimpel, H. Ludwig, A. Dan, and R. Kearney, "PANDA: Specifying Policies for Automated Negotiations of Service Contracts", *Proceedings of IEEE International Conference on Service Oriented Computing*, Rome, Italy, 2003, pp. 287-302.

[5] K.M. Sim, "G-Commerce, Market-driven G-Negotiation Agents and Grid Resource Management", *IEEE Transactions on Systems, Man and Cybernetics*, Vol. 36, No. 6, 2013, pp. 1381-1394.

[6] K.M. Sim, "A Relaxed Criteria Bargaining Protocol for Grid Resource Management", *Proceedings of 6th IEEE International Symposium on Cluster Computing and Grid Workshops*, Guangdong, China, 2012, pp. 1-13.

[7] De Sarkar Ajanta, Roy, Ghosh, Mukhopadhyay, and Nandini Mukherjee, "An Adaptive Execution Scheme for Achieving Guaranteed Performance in Computational Grids", *Journal of Grid Computing*, Vol. 8, No. 1, 2010, pp. 109-131.

[8] Ghosh P, Basu K, and Das SK, "A Game Theory-Based Pricing Strategy to Support Single/Multiclass Job Allocation Schemes for Bandwidth-Constrained Distributed Computing systems", *IEEE Transactions on Parallel and Distributed Systems* , Vol. 18, No. 3, 2007, pp. 289-306.

[9] K.M. Sim and B. Shi, "Concurrent Negotiation and Coordination for Controlling Grid Resource Co-Allocation", *IEEE Transactions on Systems, Man and Cybernetics*, Vol. 40, No. 2, 2010, pp. 753-766.

[10] K.M. Sim, "A Market-Driven Model for Designing Negotiation Agents",
     Journal of Computational Intelligence , Vol. 18, No. 4, 2002, pp. 618-637.

5

Infrastructure-as-a-Service (IaaS) Delivery Model in Cloud

Learning Objectives

After reading this chapter, you will be able to

- Describe the need for Infrastructure-as-a-Service (IaaS)

- Explain the applications of IaaS

- Compare and contrast IaaS services from several Cloud service providers

- Identify Challenges of IaaS

Massive data centers and server farms comprise the Cloud's underlying infrastructure on which Software-as-a-Service (SaaS) and Platform-as-a-Service (PaaS) run. It is an emerging computing paradigm where data and services reside in massively scalable data centers and can be ubiquitously accessed from any connected devices over the Internet. The capability provided to the consumer is to provision processing, storage, networks, and other fundamental computing resources where the consumer is able to deploy and run arbitrary software, which can include operating systems and applications.

Preliminaries The following are some of the terms and terminologies defined here for easier understanding in the remaining part of this chapter.

Virtual Machine (VM): The VM is the basic unit of computing in the system. VMs come in two flavours: non-persistent and persistent. A non-persistent machine does not persist anything after the VM is stopped. Everything on the machine that is modified while it is running is lost. In contrast, a persistent machine is backed by permanent storage (likely a virtual disk) that continues to exist after the machine is stopped. This allows the machine to be restarted into the same state that it was in the last time it was stopped.

Virtual Disk : A virtual disk is size-configurable, permanent, block-level storage that can be mounted to a running VM. Virtual disks are capable of random I/O. Virtual disks can only be mounted to a single VM at any time, but can be mounted to multiple machines throughout its lifetime.

Geographic Region: Despite the term "virtual", the machines and disks do physically exist somewhere. The geographic region is the place where the resources powering these virtual entities physically reside.

Failure-insulated Zone: These are sub-divisions of the geographic regions. Geographic regions are useful when thinking about large-scale failures and disasters (earthquakes, explosions, hurricanes) that cover a large area. Failure-insulated zones are divisions within a geographic region that are, as much as possible, isolated from expected localized failures such as disk or power supply failure.

Archival Storage: It is long term, permanent, blob-level storage. It allows the storage and retrieval of individual blobs, but does not allow random I/O within the blobs. Archival storage is not mounted to any VM, and can be accessed by multiple VMs at the same time. It exists outside of any specific geographic region. It is considered to be completely durable, but not always available.

Figure 5.1 shows IaaS, PaaS and SaaS layer. IaaS businesses offer services such as pay-as-you-go storage, networking, virtualization etc. A PaaS vendor provides hardware and software tools over the Internet, and people use these tools to develop applications. SaaS platforms make software available to users over the Internet, usually for a monthly subscription fee (e.g, email application, Customer Relationship Management (CRM) etc.). Figure 5.2 shows Cloud IaaS components, such as hardware, data centers, bandwidth, load balancers, virtual server space, and Cloud hosting. The consumers of Cloud services looks in for either of these components for running their applicaations in virtualized environment of the Cloud. This chapter looks at the general constituents of IaaS. The management of virtual machines and IaaS service providers are also covered.

5.1   IaaS in Cloud

To help understand the IaaS delivery model, the following sections examine some of its key characteristics, including dynamic scaling, agreed-upon service levels, renting, licensing, metering and self-service. All of these characteristics are the same in both public and private IaaS Cloud environments.

Dynamic scaling: Some level of uncertainty always exists when planning for IT resources. One of the major benefits of IaaS for companies faced with this type of uncertainty is the fact that resources can be automatically scaled up or down based on the requirements of the application. This important characteristic of IaaS is called dynamic scaling. If customers wind up needing

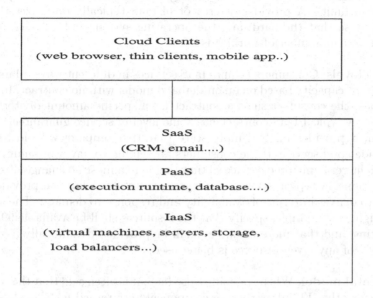

FIGURE 5.1
SaaS, PaaS, and IaaS

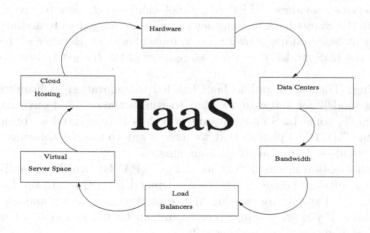

FIGURE 5.2
Components of an IaaS

more resources than expected, they can get them immediately (probably up to a given limit). A provider or creator of IaaS typically optimizes the environment so that the hardware, the operating system and automation can support a huge number of workloads.

Service levels: Consumers acquire IaaS services in different ways. Many consumers rent capacity based on an on-demand model with no contract. In other situations, the consumer signs a contract for a specific amount of storage or compute. A typical IaaS contract has some level of service guarantee. At the low end, a provider might simply state that the company will do its best to provide good service. If the consumers are willing to pay a premium price, they might get a mirrored service so that there are almost no change-of-service interruptions. A typical service-level agreement states what the provider has agreed to deliver in terms of availability and response to demand. The service level might, for example, specify that the resources shall be available 99.999% of the time and that more resources shall be provided dynamically if greater than 80% of any given resource is being used.

The rental model: When companies use IaaS, it is often said that the servers, storage, or other IT infrastructure components are rented for a fee based on the quantity of resources used and how long they are in use. Although this is true, there are some important differences between this rental arrangement and the traditional rental models that we are familiar with. For example, when we purchase server and storage resources using IaaS services, we gain immediate virtual access to the resources we need.

Within a private IaaS model, renting takes on a different focus. Although Cloud service providers (CSPs) might not charge each user to access a resource, in the charge-back model, they can allocate usage fees to an individual department based on usage over a week, month, or year. Because of the flexibility of the IaaS model, they can charge more of the budget to heavy users.

Licensing: The use of public IaaS has led to innovation in licensing and payment models for software we want to run in these Cloud environments. For example, some IaaS and software providers have created a "bring your own license" (BYOL) plan so that we have a way to use our software license in both traditional and Cloud environments.

Another option is called "pay as you go" (PAYG), which generally integrates the software licenses with the on-demand infrastructure services. For example, say that you are running Microsoft Windows Server and using the PAYG route. If you are paying ten cents an hour for Cloud access, a few cents of that fee might be going to Microsoft.

Metering and costs: We derive a potential economic benefit by controlling the amount of resources we demand and pay for it, so that we get the right match for our requirements. To ensure that users are charged for the resources they request and use, IaaS providers need a consistent and predictable way to measure usage. This process is called metering. Ideally, the IaaS provider will have a transparent process for identifying charges incurred by the user.

With multiple users accessing resources from the same environment, the IaaS provider needs an accurate method for measuring the physical use of resources to make sure each customer is charged the right amount. In addition to the basic per-instance charge discussed below, Amazon EC2 IaaS provider includes charges such as the following (note that charges can fluctuate depending on market scenarios):

1. Storage: A per-gigabyte (GB) charge for the persistent storage of data of approximately $0.00015 an hour or $0.10 a month.

2. Data transfer: A per-GB charge for data transfers of approximately $0.15. This fee may drop to $0.05 per GB, if we move large amounts of data during the billing month. Some providers offer free inbound data transfers or free transfers between separate instances on the same provider.

3. Optional services: Charges for services such as reserved IP addresses, virtual private network (VPN), advanced monitoring capabilities, or support services. Because of the variety of costs, many Cloud providers offer cost calculators, so that organizations can estimate their monthly IaaS costs. The typical billing options are as follows:

   • Pay as you go: Users are billed for resources used based on instance pricing.

   • Reserved pricing: Users are billed an upfront fee, depending on the instance type, and then billed at a discounted hourly usage rate. For some companies, this provides the best rates.

   • Trial use: Some IaaS providers, such as Amazon, offer free usage tiers that allow users to "try before they buy".

5.2 IaaS Benefits

IaaS is a good solution to the problem in many ways. The top five reasons for using IaaS are: cost savings on hardware and infrastructure, capacity management, disaster recovery/business continuity, cost savings on IT staffing and administration; and the ability to access new skills and capabilities. IaaS is frequently used for the following scenarios.

1. Organisation infrastructure: IaaS can include private and public networks. Private networks are only available to your organisation,

while public networks are shared with the world. This allows organisations to utilise pooled networking and server resources, which they
can use to run applications and store data for day-to-day business.

2. Cloud hosting: It gives a pool of virtualised workers. These can be
   utilized for sites including Internet-based applications, which exist
   on various servers dissipated over a huge topographically scattered
   network.

IaaS allows organisations to create scalable, cost-effective IT solutions
without the complexity and costly hardware required to do so in-house. IaaS
provides lower upfront cost, predictable cost, simple scalability, increased reliability and improved physical security.

5.3   Management of VMs in IaaS

There are lots of virtualization management tools, but which ones are right
for your infrastructure needs to be correctly decided. The steps involved in
managing VM are discussed here.

Creating and managing VMs

The servers and VMs tab is used to create and manage VMs. VMS can be
created using:

- ISO files in a repository (hardware virtualized only)

- Mounted ISO files on a Network File System (NFS), HTTP or FTP server
  (paravirtualized only)

- VM templates (by cloning a template)

- Existing VM (by cloning it)

- VM assemblies

VM require most installation resources to be located in the storage repository with the exception of mounted ISO files for paravirtualized guests. Before you create a VM that requires network connectivity, or a paravirtualized
machine which requires network connectivity to perform the operating system install, virtual network interfaces using the virtualized Network Interface
Card (VNIC) Manager must be generated.

The following list provides an outline of actions that can be performed in
the VM tab:

View VM information and events: You can view information and
access VM events by viewing information and events for VMs.

Edit a VM: You can edit VM parameters and you can also convert a VM to use paravirtualization. This involves making changes to the VM itself. Also start, stop, kill, restart, suspend and resuming a VM process is possible.

Move a VM between repositories: It is possible to create alternate repositories using external storage. If you have chosen to create an alternate repository, this function can be used to move a VM from one repository to another. It is important to understand that VM hosted on alternate repositories are outside of the scope of your support agreement. Moving a VM from one server to another, to or from the Unassigned VM folder can be the next process.

Migrate a VM: Migration of VMs can only be achieved between servers and between a server and the Unassigned VM folder. All compute nodes use the same CPU architecture, so you do not need to modify Server Processor Compatibility Groups.

5.4 IaaS Providers

IaaS providers maintain the equipment, while users rent or buy the specific services that they need. The term for rental or purchase is usually "pay as you go" which offers a lot of flexibility and ease in growth. Organizations use the IaaS Cloud as a way of easily scaling new or existing applications to larger audiences. Some of the commonly known service providers are discussed here.

5.4.1 Amazon AWS

Amazon Elastic Compute Cloud (Amazon EC2) is an Amazon Web Service (AWS) you can use to access servers, software, and storage resources across the Internet in a self-service manner. It provides scalable, pay as-you-go compute capacity. Also the resources provided are elastic and scales in both direction. Chief characteristics of the services provided are: (i) Amazon Machine Image (AMI) and Instance, (ii) Region & Zones, (iii) Storage, (iv) Networking and Security, (v) Monitoring, (vi) Auto Scaling and (vii) Load Balancer.

5.4.2 Windows Azure

Windows Azure started as a platform as a service (PaaS) provider but now offers Infrastructure-as-a-Service (IaaS) as well. In spite of its name, this is not a Windows-only IaaS; you can just as easily run Linux VM as would Windows Server instances. The compute and storage services in Windows Azure are typical of what you will find in most IaaS providers but with the ease that you can create a SQL Server database reminds us we are on a Microsoft platform. You also have ready access to virtual networks, service buses and non-relational storage platforms as well.

Getting Started with Windows Azure

Microsoft has one of the better IaaS management interfaces with a high-level view of all your active resources, wizards to walk you through multi-step processes, and a log display that is there when you need it but does not get in your way most of the time. Chief characteristics of Windows Azure are scalability, data protection, running applications from anywhere, and smart decision-making.

5.4.3 Google Compute Engine

Google Compute Engine lets you create and run VMs on Google infrastructure. Compute Engine offers scale, performance, and value that allows you to easily launch large compute clusters on Google's infrastructure. There are no upfront investments and you can run thousands of virtual CPUs on a system that has been designed to be fast, and to offer strong consistency of performance.

Google Compute Engine Pricing: Google Compute Engine charges for usage based on the following price sheet. A bill is sent out at the end of each billing cycle, listing previous usage and charges. Prices on this page are listed in US dollars (USD). Disk size, machine type memory, and network usage are calculated in gigabytes (GB), where 1 GB is 1073741824 bytes. This unit of measurement is also known as a gibibyte (GiB). When charging in local currency, Google will convert the prices listed into applicable local currency pursuant to the conversion rates published by leading financial institutions.

Machine type pricing: Google Compute Engine offers two categories of machine types: predefined machine types and custom machine types. Predefined machine types have preset virtualized hardware properties and a set price, while custom machine types are priced according to the number of vCPUs and memory that the VM instance uses.

Machine type billing model: The following billing model applies to all machine types, predefined or custom. All machine types are charged a minimum of 10 minutes. For example, if you run your VM for 2 minutes, you will be billed for 10 minutes of usage. After 10 minutes, instances are charged in 1 minute increments, rounded up to the nearest minute. An instance that lives for 11.25 minutes will be charged for 12 minutes of usage.

If you run an instance for a significant portion of the billing month, you can qualify for a sustained use discount. When you use an instance for more than 25% of a month, Compute Engine automatically gives you a discount for every incremental minute you use for that instance. The discount increases with usage and you can get up to a 30% net discount for instances that run the entire month. Sustained use discounts are calculated and applied to your bill as your project earns them.

Sustained use discounts for predefined machine types: When Compute Engine calculates use for predefined machine types, equivalently provisioned machines that run non-concurrently are treated as inferred instances. This gives you the flexibility to start and stop instances freely and still receive the maximum sustained use discount that is available across all of your instances. For example, if you run an n1-standard-1 instance for 75% of the month, your charges are calculated as, (i) first 25% is charged at the full on-demand rate, (ii) next 25% is charged at a 20% discount off the on-demand rate and (iii) last 25% is charged at a 40% discount off the on-demand rate.

Inferred instances: When computing sustained use discounts for predefined machine types, Compute Engine gives customer's the maximum available discount using inferred instances. An inferred instance combines multiple, non-overlapping instances in the same zone into a single instance for billing. With inferred usage, users are more likely to qualify for sustained use discounts. Non-overlapping instances running the same predefined machine type in the same zone are combined to create one or more inferred instances.

5.4.4   Rackspace Open Cloud

Designing a custom architecture for customer's mission-critical applications and workloads might seem complex, but it has solution specialists who can help customers every step of the way. From server, network and storage configuration, monitoring and support, to extending to the Cloud of your choice, they help making sure customers have everything their business needs. The customers have the service request with regard to: (i) Choose a base configuration from portfolio of dedicated server bundles to start tuning the hardware to meet their workload requirements, (ii) Chat with one of company's server specialists or engineers to customize the hardware for specific solutions, from virtualization to web applications, (iii) Rackspace deploys customer's server(s) in their data centers and (iv) Customize the application resources and the environment by leveraging Rackspace technical experts.

5.4.5   HP Enterprise Converged Infrastructure

The customers can gain more flexibility, more visibility and more freedom with the latest release of HPE Helion CloudSystem, featuring proven innovations in Cloud infrastructure, application platforms, management and hardware. The customer's business will benefit from: (i) enhanced networking, storage, servers, lifecycle management, security features, (ii) powerful tools to build and deploy apps across multiple Clouds, (iii) faster, easier, and more flexible IT infrastructure management and (iv) updated marketplace and service design.

5.5   Keys to Successfully Moving to IaaS

Many organisations have successfully implemented IaaS. While Amazon is currently the top IaaS provider, other companies such as IBM, Microsoft and more specialised providers like Rackspace and RightScale all have their place in the field. When it comes to moving to IaaS environment, there are five main things to consider, such as, assessment of current workload and processes, selection of a service that provides data sovereignty, review of SLAs, adequate attention with regard to storage capabilities, network bandwidth and compute power, and ensuring financial stability.

5.6   Challenges of IaaS

Before signing an IaaS contract, we should check the IaaS provider's information. The classic use case for Governance in Cloud Computing is when an organization wants to prevent rogue employees from misusing a service. For example, the organization may want to ensure that a user working in sales can only access specific leads and does not have access to other restricted areas. Another example is that an organization may wish to control how many virtual machines can be spun up by employees, and, indeed, that those same machines are spun down later when they are no longer needed.

We must also evaluate our regulatory compliance requirements. Usage of Cloud Services is on a paid-for basis, which means that the finance department will want to keep a record of how the service is being used. The Cloud Service Providers themselves provide this information, but in the case of a dispute, it is important to have an independent audit trail. Audit trails provide valuable information about how an organization's employees are interacting with specific Cloud services, legitimately or otherwise. The end-user organization could consider a Cloud Service Broker (CSB) solution as a means to create an independent audit trail of its Cloud service consumption.

The data center's location can play a big role in service quality and operation. A data center close to an area that frequently gets natural disaster damage may have down time. It is a good idea to research past clientele to check and see if the provider has the certifications to back up its claims. Vendor lock-in is perceived as a significant challenge to the adoption of Cloud computing. Vendors can currently mitigate this concern by making their services more transparent so that customers can understand how their resources are being managed.

Summary

In this chapter, we focused on IaaS, that is the rental of shared and virtualised computing processing, memory, and data storage for the purpose of running an operating system such as Microsoft Windows Server or Linux. This is in contrast to the hosted software application side of Cloud computing, generally known as SaaS and PaaS, discussed in the next chapters.

Keywords

IaaS	Scalability
Resiliency	Virtualization

Objective type questions

1. What second programming language did Google add for App Engine development?

(a) C++   (b) Flash

(c) Java   (d) Visual Basic

2. What facet of Cloud computing helps to guard against downtime and determines costs?

(a) Service-level agreements   (b) Application programming interfaces

(c) Virtual private networks   (d) Bandwidth fees

3. Which of these is not a major type of Cloud computing usage?

(a) Hardware-as-a-Service   (b) Platform-as-a-Service

(c) Software-as-a-Service   (d) Infrastructure-as-a-Service

4. Cloud Services have a _____ relationship with their customers.

(a) Many-to-many

(b) One-to-many

(c) One-to-one

(d) None of the above

5. What is the name of Rackspace's Cloud service?

(a) Cloud On-Demand

(b) Cloud Servers

(c) EC2

(d) None of the above

6. What is the name of the organization helping to foster security standards for Cloud computing?

(a) Cloud Security Standards Working Group

(b) Cloud Security Alliance

(c) Cloud Security WatchDog

(d) Security in the Cloud Alliance

7. Which of these companies specializes in Cloud computing management tools and services?

(a) RightScale

(b) Google

(c) Salesforce.com

(d) Savis

8. What's the most popular use case for public Cloud computing today?

(a) Test and development

(b) Website hosting

(c) Disaster recovery

(d) Business analytics

9. Virtual Machine Ware (VMware) is an example of

(a) Infrastructure Service

(b) Platform Service

(c) Software Service

(d) All of the above

10. Cloud Service consists of

(a) Platform, Software, Infrastructure

(b) Software, Hardware, Infrastructure

(c) Platform, Hardware, Infrastructure

(d) None of the above

11. Which of the following can be considered PaaS offering ?

(a) Google Maps

(b) Gmail

(c) Google Earth

(d) All of the above

12. _____ is the most refined and restrictive service model.

(a) IaaS

(b) CaaS

(c) PaaS

(d) All of the above

13. _____ provides virtual machines, virtual storage, virtual infrastructure and other hardware assets.

(a) IaaS

(b) SaaS

(c) PaaS

(d) All of the above

14. Amazon Web Services offers a classic Service Oriented Architecture (SOA) approach to :

(a) IaaS

(b) SaaS

(c) PaaS

(d) All of the above

15. Rackspace Cloud Service is an example of :

(a) IaaS

(b) SaaS

(c) PaaS

(d) All of the above

Objective type questions -answer

1:c   2:a   3:a   4:b   5:b   6:a   7:b   8:a   9:a   10:a   11:a   12:c   13:a   14:c   15:a

Review questions

1. Define IT infrastructure from both a technology and a services perspective.

2. List each of the eras in IT infrastructure evolution and describe its distinguishing characteristics.

3. Define and describe the following: Web server, application server, multi-tiered client/server architecture.

4. Describe how network economics, declining communications costs, and technology standards affect IT infrastructure.

5. What are the stages and technology drivers of IT infrastructure evolution?

6. List and describe the components of IT infrastructure that firms need to manage.
7. Describe the evolving mobile platform, quantum computing and Cloud computing.
8. Explain how businesses can benefit from virtualization, green computing and multicore processors.
9. What are the challenges of managing IT infrastructure and management solutions?
10. How is IaaS different from SaaS and PaaS?

Critical thinking questions

1. Analyze the benefits and the problems posed by the four approaches for the implementation of resource management policies: control theory, machine learning, utility-based, market-oriented.
2. Creating a virtual machine (VM) reduces ultimately to copying a file, therefore the explosion of the number of VMs cannot be prevented. As each VM needs its own IP address, virtualization could drastically lead to an exhaustion of the $IPv4$ address space. Analyze the solution to this potential problem adopted by the IaaS Cloud service delivery model.
3. Analyze the reasons for the introduction of storage area networks (SANs) and their properties.
4. In a scale-free network the degree of the nodes have an exponential distribution. A scale-free network could be used as a virtual network infrastructure for Cloud computing. Controllers represent a dedicated class of nodes tasked with resource management; in a scale-free network nodes with a high connectivity can be designated as controllers. Analyze the potential benefit of such a strategy.

Bibliography

[1] Xabriel J, Collazo-Mojica Jorge Ejarque, Masoud Sadjadi S, Rosa M Badia, "Cloud application resource mapping and scaling based on monitoring of QoS constraints", *Proceedings of the 2012 International Conference on software engineering and knowledge engineering*, Vol. 7, No. 4, 2012, pp. 88-93.

[2] Xie J, "Improving map reduce performance through data placement in heterogenous Hadoop clusters", *IEEE Int Symp. Parallel Distributed Process 2010*, Vol. 66, No. 10, pp. 1322-1337.

[3] Yazir YO, Matthews C, Farahbod R, Neville S, Guitouni A, Ganti S, Coady Y, "Dynamic resource allocation based on distributed multiple criteria decisions in computing Cloud", *Proceedings of the 3rd International conference on Cloud computing*, Vol. 28, No. 1, 2010, pp. 91-80.

[4] Yoshino T, Osana Y, Kuribayashi S, "Evaluation of congestion control methods for joint multiple resource allocation", *Proceeding of the 13th International Conference on Network-based Information Systems*, Vol. 10, No. 3, 2010, pp. 16-28.

[5] Younge AJ, von Laszewski G, Wang L, Lopez-Alarcon S, Carithers W, "Efficient resource management for Cloud computing environments", *Proceedings of the International Conference on Green Computing*, Vol. 50, No. 12, 2011, pp. 60-71.

[6] Zhang Q, Quanyan Zhu, Raouf Boutaba, "Dynamic resource allocation for spot markets in Cloud computing environment", *Proceedings of the 4th IEEE International Conference on Utility and Cloud computing*, Vol.10, No. 6, 2011, pp. 177-185.

[7] Zhang S, Zhuzhong Qian, Jie Wu, "An opportunistic resource sharing and topologyaware mapping framework for virtual networks", *Proceedings of the IEEE INFOCOM 2012*, Vol. 13, No. 4, 2012, pp. 2408-2416.

[8] Zhu Q, Gagan Agrawal, "Resource provisioning with budget constraints for adaptive applications in Cloud environments", *Proceedings of the HPDC 2010*, Vol. 8, No. 3, 2010, pp. 304-307.

6

SaaS and PaaS in Cloud

Learning Objectives

After reading this chapter, you will be able to

- Define the characteristics of SaaS and PaaS

- Analyze the advantages/disadvantages of SaaS and PaaS

- Compare and contrast implementation/ examples w.r.t SaaS and PaaS

This chapter helps to develop an understanding of the high-level differences between SaaS and PaaS. Before the customer has made the decision to consider Cloud services for their application or infrastructure deployment, it is important that they understand the fundamental differences between the core categories of Cloud services available.

6.1 SaaS in Cloud

The idea of using SaaS first popped up in the late 1990s in order to allow sharing end-user licenses in a way that reduced cost and also shifted infrastructure demands from the company to the software provider. SaaS is a software distribution model in which a third-party provider hosts applications and makes them available to customers over the Internet.

Adobe Photoshop Express, Online image processing, fluidOps, eCloud Manager, SAP Edition, SAP Landscape as a Service, Google Docs, Online office applications, Google Google Maps API, Service for the integration of maps and geo information, Google Open Social Generic programming interface for the integration of social networks into applications OpenID, Distributed cross-system user identity management system, Microsoft Windows Live Online office applications, Salesforce.com Extensible Customer Relationship Management (CRM) system are some of the examples of SaaS (Figure 6.1).

Preliminaries

Some of the terms and terminologies defined here for easier understanding in the remaining part of this chapter.

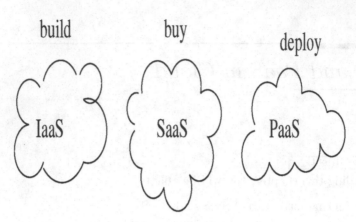

FIGURE 6.1
SaaS delivery model

Content Delivery Network (CDN) : It is basically a system of distributed servers which enables you to serve content to your app users with high performance and high availability.

Application Programming Interface (API): An API is an interface that allows the user to access information from another service and integrate this service into their own application. Through a set of defined requests, the asking application is allowed to access limited pieces of the called upon application's functionality. APIs are used to share limited functionality between programs. One example of an API is the Facebook share button on this page, another is Yelp's use of Google maps to display nearby restaurants.

MongoDB): MongoDB is a cross-platform document-oriented database program. MongoDB is great for transactional stores where performance is a concern. It is also great when the data structure is going to evolve over time, as its schema-less operations allow you to update the data on the fly.

Stackato): Stackato is an application platform for creating your own private Platform as a Service (PaaS) using any language on any stack in any Cloud.

Web 2.0: Web 2.0 refers to websites that emphasize user-generated content, ease of use, participatory culture and interoperability for end users.

6.2 SaaS Characteristics

Although not all SaaS applications share all traits, the characteristics below are common among many SaaS applications.

Configuration and customization: SaaS applications support what is traditionally known as application customization. In other words, like traditional enterprise software, a single customer can alter the set of configuration options (parameters) that affect its functionality and look-and-feel. Each customer may have its own settings (parameter values) for the configuration options. The application can be customized to the degree it has been designed for based on a set of predefined configuration options. For example, to support customer's common need to change an application's look-and-feel so that the application appears to be having the customer's brand, many SaaS applications let customers provide (through a self service interface or by working with application provider staff) a custom logo and sometimes a set of custom colors. The customer cannot, however, change the page layout unless such an option was designed for.

Accelerated feature delivery: SaaS applications are often updated more frequently than traditional software, in many cases on a weekly or monthly basis. This is enabled by several factors. If the application is hosted centrally, an update is decided and executed by the provider, not by customers. When the application only has a single configuration, making development testing faster. The application vendor does not have to expend resources updating and maintaining backdated versions of the software, when there is only a single version. The application vendor has access to all customer data, expediting design and regression testing. The solution provider has access to user behavior within the application (usually via web analytics), making it easier to identify areas worthy of improvement. Accelerated feature delivery is further enabled by agile software development methodologies. Such methodologies, which have evolved in the mid-1990s, provide a set of software development tools and practices to support frequent software releases.

Open integration protocols: Because SaaS applications cannot access a company's internal systems (databases or internal services), they predominantly offer integration protocols and application programming interfaces (APIs) that operate over a wide area network. Typically, these are protocols based on HTTP, REST and SOAP (see chapter 2). The ubiquity of SaaS applications and other Internet services and the standardization of their API technology has spawned development of mashups, which are lightweight applications that combine data, presentation and functionality from multiple services, creating a compound service. Mashups further differentiate SaaS applications from on-premises software as the latter cannot be easily integrated outside a company's firewall.

Collaborative (and "social") functionality: Inspired by the success of online social networks and other so-called web 2.0 functionality, many SaaS applications offer features that let its users collaborate and share information. For example, many project management applications delivered in the SaaS model offer-in addition to traditional project planning functionality-collaboration features letting users comment on tasks and plans and share documents within and outside an organization. Several other SaaS applications let users vote on and offer new feature ideas. Although some collaboration-related functionality is also integrated into on-premises software, implicit or explicit collaboration between users or different customers is only possible with centrally hosted software.

6.3   SaaS Implementation

SaaS implementation refers to the tasks that must be completed to successfully enable a SaaS offering in a Cloud computing environment. SaaS enablement can occur through a self service provisioning system for simple applications that are typically made available via public Clouds. On the other hand, SaaS offerings which are designed for private enterprise use frequently require hands on configuration that must be undertaken by the independent software vendor (ISV) for SaaS enablement to occur.

Here, we shall see on how to start building a Cloud-based SaaS architecture, dealing with issues of scalability and what this means for your SaaS application. When building a (global) SaaS application, there are chances that you are building it in the Cloud. The Cloud has a lot of advantages, such as scalability in contrast to local server environments. Other advantages are presented as follows.

Programming language to use: Building a product for the Cloud means building a product with a modern programming language. Besides personal abilities and skills, the choice of your programming language will be influenced by the possibilities of each language. There are various (modern) programming languages out there making it a hard time to choose the right one. We need to choose the most prominent ones, play around with those and try to experiment as much as possible. Python is a widely used programming language, designed to emphasize on its code readability.

Perfect database selection: Another imprtant thing includes the installation of a database. Use of a document-oriented database (e.g. MongoDB) is encouraged. Document databases are quite different to the traditional concept of relational databases. Document databases get their type information from the data itself. Thus, every instance of data can be different from any other. This allows more flexibility, especially when dealing with changes. And it often reduces database sizes.

Queuing system: A message queuing system is an asynchronous communication protocol, enabling sender and receiver of a message not interacting at the same time. Also known as Message Queuing (MSMQ) technology, it enables web apps to run at different times and to communicate with various 3rd party integrations /APIs/ and other services asynchronously.

Web Storage S3: Getting more and more users on board for your product will make you easily wonder about your web storage. With the Amazon S3 storage service, a highly scalable object storage can be installed.

Content Delivery Network (CDN): It is basically a system of distributed servers which enables you to serve content to your app users with high performance and high availability. Let's assume you have 3 EC2s installed. One in the US, one in Europe and one in Singapore. If someone from New York visits your app, the CDN enables you to serve content to the user through the EC2 located in the US.

A SaaS implementation in a private Cloud can be simple or complex, depending on the scope of the various elements to be included in the configuration, such as: business processes, workflows, set-up data, user roles and responsibilities, meta data; and custom instructions and error messaging. Incidentally, ISV's can increase their profit margins when implementing SaaS solutions, through the creation of configuration templates that can be reused to configure common elements across multiple implementations.

6.4   SaaS Examples

Here we discuss some successful SaaS businesses and the services they provide.

HubSpot: HubSpot is, at its core, a client/customer relationship management (CRM) system. A CRM is typically used to manage the sales process, moving somebody from a lead to a prospect, and finally to a customer. It allows you to manage the information related to their relationship with the business. HubSpot provides a free forever tier (often referred to as "freemium") for their CRM with the goal to get potential customers to start using the basic software and then pay to upgrade to the integrated and advanced outbound marketing tools offered. It provides different pricing tiers that scale with your

business and is considered to have one of the best systems out there for on-
boarding new customers.

Skubana: Skubana provides an integrated inventory management solution
for online retailers who want to sell their products via multiple distribution
channels. High-quality inventory management software can be expensive, and
a centralized inventory is critical to ensure a retailer does not oversell their
stock, which is a big problem when it comes to managing multiple sales channels. Skubana provides a Cloud-based solution at a fraction of the cost of
implementing a custom inventory management system from scratch.

Shopify: Setting up an online store yourself can be complicated and take outside expertise from developers and designers to make it work. Shopify creates a
platform where you could build an ecommerce store and start selling products
in just a few hours. Setting up a Shopify store is easy and painless, providing
the site, the shopping cart, online catalog, and payment integration all in the
one platform. It also offers customization options via its library of both free
and paid plugins. If you are dropshipping or looking to run a branded site for
your business that integrates with your Fulfillment by Amazon business, then
Shopify offers a complete and simple solution with competitive pricing.

G Suite: Google's G Suite is basically a Cloud version of Microsoft Office,
albeit with a little less functionality. G Suite is officially the name for Google's
business version of apps; however, the paid version is only a fraction of the
price you would pay for Microsoft Office. If you have a free Gmail account,
Google Apps is the free version of the same software. Many business owners
never need to use anything else to run their businesses. It also has hands down
the best collaboration functionality of all the online business apps, where you
can have multiple users logged into the same document and making changes
simultaneously without creating versioning issues like some other services.

Zendesk: Zendesk provides a centralized way to manage all of your customer
support tickets. Instead of filling up some support inbox or losing emails over
several different customer support team inboxes, Zendesk provides a way to
track, allocate, communicate and resolve any issues of customers.

Dropbox: Dropbox was one of the first consumer-grade Cloud storage solutions on the web. Dropbox has several pricing tiers, starting from their free
plan (beginning with 2 GB of storage) for individuals, and 2 TB or more
space on their business offerings. The basic levels offer an easy way to save
your photos and files, and sync them across your devices regardless of the OS
you are running. At the higher end, it allows for you to recover deleted data
and offers encryption options with advanced user permission management,
much like what you would find on an in-house corporate network — without
the need to invest in expensive software and infrastructure.

Slack: The default communication medium for most businesses is either phone or email; however, if you are running a business with a remote team, you need something a little more flexible. Slack is an instant messaging service that gives you the convenience of Facebook Messenger, but with advanced functionality. It also allows easy collaboration with multiple users and the ability to share sensitive company data securely. It also supports integrations with most popular project management apps, such as Asana, Trello and Basecamp. Slack offers a freemium tier to try it out as well as a paid version.

Adobe Creative Cloud: Adobe has a long history and is the industry standard for anyone doing graphic design or video production. Adobe used the perpetual license model for many years, but a single piece of their software suite would typically sell for over $1,500 — putting it out of reach for many small business owners. This may be the reason it used to be among the most pirated software on the web. Adobe saw the light and shifted to a SaaS subscription model in 2013, supplying their entire suite of products for only $49.99 a month — a fraction of what it used to cost to buy a single piece of their software. This allowed them to scale up and increase their user base by improving accessibility to their suite of products through more competitive pricing.

Microsoft Office 365: Microsoft Office 365 is the industry standard enterprise-level software for larger corporations — offering more advanced features than Google's service. Microsoft has been around a long time, and many people are using it. Office 365 can provide the benefits of the Cloud while still offering something familiar for your team, reducing the time and cost. Google Apps are also quite limited in their functionality compared to Office 365. This is most noticable if you do any advanced spreadsheet work, where Excel wins hands down. Another key point is Office 365 will keep working if you lose your internet connection, which is important for many business owners who may need to access their data even if they are offline – or simply want to keep their data off Google's servers.

6.5 Advantages and Disadvantages of SaaS

The benefits of the SaaS model are as follows.

Flexible payments: As opposed to buying software to install, or extra hardware to run the applications, clients buy in to a SaaS offering. For the most part, they pay for this services on a month to month premise utilizing a pay-as-you-go. Progressing expenses to a repetitive working cost permits numerous organizations to practice better and more unsurprising planning. Clients can likewise end SaaS contributions whenever to stop those common expenses.

Scalable usage: Cloud services like SaaS offer high scalability, which gives customers the option to access more, or fewer, services or features on-demand.

Automatic updates: Rather than purchasing new software, customers can rely on a SaaS provider to automatically perform updates and patch management. This further reduces the burden on in-house IT staff.

Accessibility and persistence: Since SaaS applications are delivered over the Internet, users can access them from any Internet-enabled device and location.

SaaS likewise represents some expected drawbacks. Organizations must depend on outside vendors to provide the software, keep that product fully operational, track and report exact charging and encourage a secure environment for the business information. Suppliers that experience administration interruptions, force undesirable changes to support contributions, experience a security penetrate or some other issue can profoundly affect the client's capacity to utilize those SaaS contributions. Subsequently, clients ought to comprehend their SaaS supplier's administration level understanding, and ensure that it is implemented.

SaaS is firmly identified with the Application Service Provider (ASP) and on demand computing software delivery models. The provider provisions the software to the clients over the Internet. Organizations can integrate SaaS applications with other software using application programming interfaces (APIs). For example, a business can write its own software tools and use the SaaS provider's APIs to integrate those tools with the SaaS offering. There are SaaS applications for fundamental business technologies, such as email, sales management, customer relationship management (CRM), financial management, human resource management, billing and collaboration. Leading SaaS providers include Salesforce, Oracle, SAP, Intuit and Microsoft.

Solution providers should be evaluated on two main dimensions, (i) SaaS delivery: Since SaaS is an out-tasking model, it is important to verify a provider can deliver the level of service and capabilities the company requires, and (ii) Functionality: The customer's need to be sure that the service will meet the document management and fulfill their demands and will support the security, compliance and cost considerations of the business.

6.6   PaaS in Cloud

PaaS is a category of Cloud computing services that provides a platform allowing customers to develop, run, and manage applications without the complex-

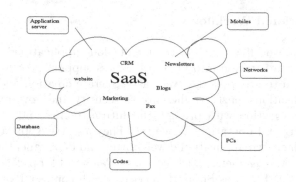

FIGURE 6.2
PaaS delivery model

ity of building and maintaining the infrastructure typically associated with developing and launching an application. As a layman, we can use the word PaaS and Middleware interchangeably as the use case for both of them is same i.e., Custom Application Development. Middleware is software that lies between an operating system and the applications running on it. Essentially functioning as hidden translation layer, middleware enables communication and data management for distributed applications (Figure 6.2). PaaS can be delivered in two ways: as a public Cloud service from a provider, where the consumer controls software deployment with minimal configuration options, and the provider provides the networks, servers, storage, OS, "middleware" (i.e.; java runtime, .net runtime, integration, etc.), database and other services to host the consumer's application; or as a private service (software or appliance) inside the firewall, or as software deployed on a public IaaS.

The original intent of PaaS was to simplify the code-writing process for developers, with the infrastructure and operations handled by the PaaS provider. Originally, all PaaSes were in the public Cloud. Because many companies did not want to have everything in the public Cloud, private and hybrid PaaS options (managed by internal IT departments) were created.

PaaS offerings may also include facilities for application design, application development, testing and deployment, as well as services such as team collaboration, web service integration, and marshalling, database integration, security, scalability, storage, persistence, state management, application versioning, application instrumentation, and developer community facilitation. Besides the service engineering aspects, PaaS offerings include mechanisms for service management, such as monitoring, workflow management, discovery and reservation.

PaaS Operational Workflows

Operational workflows enable creation of Cloud applications and PaaS services and their integration into IT and business support processes. Four categories are used to group workflows that accelerate creation and deployment of Cloud applications, ensure adherence to security and governance policies, standardize integration with present and future tooling and link cohesively into monitoring and operations processes. They are: (i) Application Support Services which govern application development and deployment activities, (ii) Cloud Service Management which connects PaaS to IT operations and support process, (iii) Business Support Services which connect PaaS to business processes managed outside of IT, and (iv) Cloud Services Development which streamline and standardize the creation of services.

Within these four areas, consideration should be given to the following areas to support the Private PaaS.

Application Delivery Lifecycle Workflows The purpose of PaaS is to eliminate manual configuration of the systems and software required for running an application and hosting its data. It must expose a self-service interface for deploying applications so that developers are free to focus on application code rather than configuration management and system setup. Cloud service providers enables developers to deploy, update, manage and monitor applications. Developers push applications to each environment, using a command line client or IDE, along with basic configuration information. Cloud service providers takes care of the rest, including configuring an appropriate web server, providing an application URL, allocating memory, providing the appropriate language runtime, provisioning a database for the application, installing supplemental language modules, starting the requested number of application instances, round-robin routing to distribute traffic to multiple instances.

In a traditional development/deployment workflow, inconsistencies in software versions between each stage in the process often cause malfunctions or breakage in the application. Cloud service providers provide the same software components and virtual hardware at each stage eliminates these hurdles. Development flows into production deployment with less friction and less time spent addressing incompatibilities in the supporting software. All stages from development to production use Stackato, providing the same components, using the same versions, creating a consistent environment throughout regardless of the host operating system.

PaaS-Deployed Application Architecture Applications deployed, hosted and managed with Cloud service providers can include standard multi-tier applications, web applications, mobile applications, web services (apps providing services to other apps), worker process, cron jobs, or a mix of the above. A cron job is a Linux command used for scheduling tasks to be executed sometime in the future. An application may be deployed as a single monolithic, legacy application, or be comprised of several different applications representing pieces of a larger solution. Applications may choose to have fully isolated or shared data and filesystem space within a cluster, and sharing is also possible between applications deployed using different language runtimes. Applications may connect with each other over standard http or via TCP or UDP ports through the use of Stackato's harbor service.

The main roles are as follows:

- Router: directs all web application web traffic to the appropriate application containers running on DEA nodes.

- Controller: exposes the "API endpoint", runs the Management Console, and generally coordinates the operation of all other nodes and roles.

- Droplet Execution Agent (DEA): runs multiple user applications inside Ephemeral Linux containers. The advantages of containers are that they run isolated processes by providing all needed dependencies using an immutable approach. By adding only the required dependencies into the image, a container lowers attack vectors and provides faster startups and deployments.

- Service roles: a variety of services which can be bound to applications on demand (e.g. databases, cache, port forwarding, message queue, persistent filesystem)

Calculating Memory Requirements

Since most current Cloud deployment methodologies have applications deployed to their own separate VM, it's necessary to re-evaluate the memory usage and performance when deployed to a multi-tenant system using application containers. Multi-tenancy means that a single instance of the software and its supporting infrastructure serves multiple customers.

Integrating with Your IT Environment

Greater utilization by developers, and acceptance by various IT influencers will come when Private PaaS is integrated with other systems that are already in place. Some clients start with Cloud service providers in an isolated environment limited to a specific set of developers that are demanding the benefits of private PaaS. Then, they move on with integrating the most critical areas that will impact user acceptance and compliance. Customers have integrated tools with existing IT and Development systems, including existing virtualized and IaaS environments, both in the data center and public Clouds, user authentication & authorization, enterprise databases (i.e. Oracle), monitoring systems, log management systems, Continuous Integration (CI) systems, Source Code Management (SCM) systems, metering & billing systems.

6.7   PaaS Characteristics

The essential characteristics of PaaS are as follows.

Runtime Framework: This is the "software stack" aspect of PaaS, and perhaps the aspect that comes first to mind for most people. The PaaS runtime framework executes end-user code according to policies set by the application owner and Cloud provider. PaaS runtime frameworks come in many flavors, some based on traditional application runtimes, others based on Fourth Generation Ganguage (4GL) and visual programming concepts, and some with pluggable support for multiple application runtimes.

Abstraction: Platform-oriented Cloud platforms are distinguished by the higher level of abstraction they provide. With IaaS, the focus is on delivering to users "raw" access to physical or virtual infrastructure. In contrast, with PaaS, the focus is on the applications that the Cloud must support. Whereas an IaaS Cloud gives the user a bunch of virtual machines that must be configured and to which application components must be deployed, a PaaS Cloud provides the user a way to deploy her applications into a seemingly limitless pool of computing resources, eliminating the complexity of deployment and infrastructure configuration.

Automation: PaaS environments automate the process of deploying applications to infrastructure, configuring application components, provisioning and configuring supporting technology like load balancers and databases, and managing system change based on policies set by the user. While IaaS is known for its ability to shift capital costs to operational costs through outsourcing, only PaaS is able to slash costs across the development, deployment and management aspects of the application lifecycle.

Cloud Services: PaaS offerings provide developers and architects with services and APIs that help simplify the job of delivering elastically scalable, highly available Cloud applications. These Cloud services provide a wide variety of capabilities, and in many instances are key differentiators among competing PaaS offerings. Examples of Cloud services include services and APIs for distributed caching, queuing and messaging, workload management, file and data storage, user identity, analytics, and more. By providing built-in Cloud services, platform offerings eliminate the need to integrate many disparate components and decrease time-to-market for applications on the platform.

6.8 PaaS Implementation

For a successful enterprise-wide implementation of a PaaS, the Cloud service providers recommend the following approach and considerations. The successful implementation of a private PaaS starts with proper resourcing behind the initiative. The number of people behind the project depends on the size of the implementation and complexity of the organization, but a typical implementation that starts "small", then grows to an enterprise-wide implementation should be staffed with the following people and skills.

Project Owner: Every successful IT project needs a champion behind it, and private PaaS is no exception. This person is not necessarily a project manager, but is usually the person that initiates the private PaaS project because he/she really believes in the benefits it would bring to the organization and he/she will drive others on the team and organization forward.

Executive Sponsor: The larger the project and impact across functional teams, the greater the importance for a VP or C-level executive sponsor. This person doesn't necessarily need to be involved in the day-to-day implementation or running of a PaaS, but the Project Owner will report to him/her and communicate any issues and results. If the project is starting small, the Executive Sponsor will want to monitor the results in order to decide or influence what other groups in the organization it should expand to.

System Administrator: A small departmental implementation may only have one system administrator, but a larger implementation should have two administrators that are responsible for the set-up and ongoing maintenance. This person should have IT competence in virtualized infrastructure, basic Linux/UNIX system administration, and network configuration. Junior Administrators need not be experts in these domains, but should have a basic understanding of the requirements of software application and database hosting.

Architects: Architects will work with all the primary stakeholders in the roles listed here, plus others that can influence the use and running of a PaaS to ensure alignment with overall IT and development strategy. Architects should have an understanding of current infrastructure, middleware and application technologies used and how PaaS fits into the picture or displaces legacy technologies, and improves processes.

6.9   PaaS Examples

There are two prominent types of PaaS. Public PaaS is delivered by a services provider for building applications. Examples include Salesforce Heroku, AWS Elastic Beanstalk, Microsoft Azure, and Engine Yard. Enterprise PaaS is delivered by central IT within an organization to developers and possibly partners and business customers. Enterprise PaaS sits on top of public IaaS, on-premise bare metal, and on-premise virtual machines. Some technology analysts make a distinction between the actual service that central IT is delivering (PaaS) and the software used to deliver that service. For example, Gartner uses the term Cloud Enabled Application Platform (CEAP). Examples include Apprenda, VMware, EMC-owned Pivotal, and Red Hat OpenShift.

6.10   Advantages and Disadvantages of PaaS

The advantages of PaaS are primarily that it allows for higher-level programming with dramatically reduced complexity; the overall development of the application can be more effective, as it has built-in infrastructure; and maintenance and enhancement of the application is easier. It can also be useful in situations where multiple developers are working on a single project involving parties who are not located nearby.

One disadvantage of PaaS offerings is that developers may not be able to use a full range of conventional tools (e.g. relational databases, with unrestricted joins). Another possible disadvantage is being locked in to a certain platform. However, most PaaSes are relatively lock-in free.

6.10.1   Types

Public, private and hybrid

There are several types of PaaS, including public, private and hybrid. PaaS was originally intended as an application solution in the public Cloud, before expanding to include private and hybrid options.

Public PaaS is derived from software as a service (SaaS), and is situated in Cloud computing between SaaS and IaaS. SaaS is software that is hosted in the Cloud, so that it doesn't take up hard drive from the computer of the user or the servers of a company. IaaS provides virtual hardware from a provider with adjustable scalability. With IaaS, the user still has to manage the server, whereas with PaaS the server management is done by the provider. IBM Bluemix (also private and hybrid), Amazon AWS and Heroku are some of the commercial public Cloud PaaS providers.

A private PaaS can typically be downloaded and installed either on a company's on-premises infrastructure, or in a public Cloud. Once the software is installed on one or more machines, the private PaaS arranges the application and database components into a single hosting platform. Private PaaS vendors include Apprenda, which started out on the Microsoft .NET platform before rolling out a Java PaaS; Red Hat's OpenShift and Pivotal Cloud Foundry. Apprenda and Microsoft once considered to be two of the only PaaSes that provide superior .NET support . Now joined by the publicly announced Microsoft and IBM Partnership programme.

Hybrid PaaS is typically a deployment consisting of a mix of public and private deployments. An example here is IBM Bluemix which is delivered as a single, integrated Cloud platform across public, dedicated, and on-premise deployment models.

Mobile PaaS

Initiated in 2012, mobile PaaS (mPaaS) provides development capabilities for mobile app designers and developers. The Yankee Group identified mPaas as one of its themes for 2014, naming a number of providers including Kinvey, CloudMine, AnyPresence, FeedHenry, FatFractal and Point.io.

Open PaaS

Open PaaS does not include hosting, but rather it provides open source software allowing a PaaS provider to run applications in an open source environment. For example, AppScale allows a user to deploy some applications written for Google App Engine to their own servers, providing datastore access from a standard SQL or NoSQL database. Some open platforms let the developer use any programming language, database, operating system or server to deploy their applications.

PaaS for Rapid Development

In 2014, Forrester Research defined Enterprise Public Cloud Platforms for Rapid Developers as an emerging trend, naming a number of providers in cluding Mendix, Salesforce.com, OutSystems and Acquia.

Summary

This chapter discussed about different Cloud service delivery model, namely PaaS and SaaS. Overall, each Cloud model offers its own specific features and functionalities, and it is crucial for an organization to understand the differences. Whether one is looking for Cloud-based software for storage options, or are wanting complete control over their entire infrastructure without having to physically maintain it, there is a Cloud service for all.

Keywords

PaaS	IaaS
SaaS	Delivery models

Objective type questions

1. Which of the following is true of Cloud computing?

(a) It is always going to be less expensive and more secure than local computing

(b) You can access your data from any computer in the world, as long as you have an Internet connection

(c) Only a few small companies are investing in the technology, making it a risky venture

(d) None of the above

2. What is private Cloud?

(a) A standard Cloud service offered via the Internet

(b) A Cloud architecture maintained within an enterprise data center

(c) A Cloud service inaccessible to anyone but the cultural elite

(d) None of the above

3. Amazon Web Services is which type of Cloud computing distribution model?

(a) Software as a Service (SAAS)

(b) Platform as a Service (PAAS)

(c) Infrastructure as a Service (IAAS)

(d) None of the above

4. Google Docs is a type of Cloud computing

(a) True

(b) False

5. What is Cloud Foundry?

(a) A factory that produces Cloud components

(b) An industry wide PaaS initiative

(c) VMware-led open source PaaS

(d) None of the above

6. This is a software distribution model in which applications are hosted by a vendor or service provider and made available to customers over a network, typically the Internet

(a) Platform as a Service (PaaS)

(b) Infrastructure as a Service (IaaS)

(c) Software as a Service (SaaS)

(d) None of the above

7. Which of the following statements about Google App Engine (GAE) is INCORRECT.

(a) It's a Platform as a Service (PaaS) model

(b) Automatic Scalability is built in with GAE. As a developer you don't need to worry about application scalability

(c) You can decide on how many physical servers required for hosting your application

(d) The applications deployed on GAE have the same security, privacy and data protection policies as that of Google's applications. So, applications can take advantage of reliability, performance and security of Google's infrastructure

8. I've a website containing all static pages. Now I want to provide a simple Feedback form for end users. I don't have software developers, and would like to spend minimum time and money. What should I do?

(a) Hire software developers, and build dynamic page

(b) Use ZOIIO creator to build the required form, and embed in html page

(c) Use Google App Engine (GAE) to build and deploy dynamic page

(d) None of the above

9. What is the name of the organization helping to foster security standards for Cloud computing?

(a) Cloud Security Standards Working

(b) Cloud Security Alliance

(c) Cloud Security WatchDog

(d) Security in the Cloud Alliance

10. "Cloud" in Cloud computing represents what?

(a) Wireless

(b) Hard drives

(c) People

(d) Internet

11. Which of the following is a SaaS characteristic ?

(a) The typical license is subscription-based or usage-based and is billed on a recurring basis

(b) The software is available over the Internet globally through a browser on demand

(c) The software and the service are monitored and maintained by the vendor

(d) All of the mentioned

12. _____ applications have a much lower barrier to entry than their locally installed competitors.

(a) IaaS

(b) CaaS

(c) PaaS

(d) None of the above

13. SaaS supports multiple users and provides a shared data model through _____model.

(a) single-tenancy

(b) multi-tenancy

(c) multiple-instance

(d) All of the above

14. Open source software used in a SaaS is called _____ SaaS.

(a) closed

(b) free

(c) open

(d) All of the aboveg

15. The componentized nature of SaaS solutions enables many solutions to support a feature called :

(a) workspace

(b) workloads

(c) mashups

(d) All of the above

Objective type questions -answer

1:b   2:b   3:c   4:a   5:c   6:c   7:c   8:b   9:b   10:d   11:d   12:d   13:b   14:c   15:c

Review questions

1. Mention the platforms which are used for large scale Cloud computing?

2. What is the difference in Cloud computing and computing for mobiles?

3. List out different layers which define Cloud architecture?

4. What is the requirement of virtualization platform in implementing Cloud?

5. Explain what are the different modes of software as a service (SaaS)?

6. In Cloud computing what are the different layers?

7. How important is the platform as a service?

Critical thinking questions

1. Mobile devices could benefit from Cloud computing; explain the reasons you think that this statement is true or provide arguments supporting the contrary. Discuss several Cloud applications for mobile devices; explain which one of the three Cloud computing delivery models, SaaS, PaaS, or IaaS, would be used by each one of the applications and why.

2. Compare the three Cloud computing delivery models, SaaS, PaaS and IaaS, from the point of view of the application developers and users. Discuss the security and the reliability of each one of them. Analyze the differences between the PaaS and the IaaS.

3. An IT company decides to provide free access to a public Cloud dedicated to higher education. Which one of the three Cloud computing delivery models, SaaS, PaaS, or IaaS should it embrace and why? What applications would be most beneficial for the students? Will this solution have an impact on distance learning? Why?

4. Software licensing is a major problem in Cloud computing. Discuss several ideas to prevent an administrator from hijacking the authorization to use a software licence.

Bibliography

[1] J. W. J. Y. Lee, "A Quality Model for Evaluating Software-as-a-Service in Cloud Computing", *Software Engineering Research, Management and Applications*, 2009.

[2] P. X. Wen and L. Dong, "Quality Model for Evaluating SaaS Service", *Proceedings of Fourth International Conference on Emerging Intelligent Data and Web Technologies*, 2013, pp. 83-87.

[3] Gajah, S. N. R. Sankranti, "Adaptive of SERVQUAL Model in Measuring Customer Satisfaction towards Service Quality Provided by Bank Islam Malaysia Berhad (BIMB) in Malaysia", *Int. J. Bus. Soc. Sci.*, Vol. 4, No. 10, pp. 189-198, 2013.

[4] A Parasuraman, V. A Zeithaml, and L. L. Berry, "SERQUAL: A Multiple-Item scale for Measuring Consumer Perceptions of Service Quality", *Int. J. Bus. Soc. Sci.*, Vol. 64, 1988, pp. 28-29.

[5] Dhanamma Jagli, Dr. Seema Purohit, Dr N. Subhash Chandra "SAASQUAL : A Quality Model for Evaluating SaaS on The Cloud", *Int. J. Bus. Soc. Sci.*, pp. 1-6, 2015.

[6] Q. He, J. Han, Y. Yang, J. Grundy, and H. Jin, "QoS-driven service selection for multi-tenant SaaS", *Proceedings of Proc. IEEE 5th Int. Conf. Cloud Comput. Cloud*, pp. 566–573, 2012.

[7] Z. Wang, N. Jiang, and P. Zhou, "Quality Model of Maintenance Service for Cloud Computing", *Proceedings of IEEE 17th Int. Conf. High Perform. Comput. Commun. (HPCC)*, pp. 1460–1465, 2015.

[8] D. Beimborn, T. Miletzki, and D. W. I. S. Wenzel, "Platform as a service (PaaS)", *Wirt schafts informatik*, Vol. 53, No. 6, pp. 371-375, 2011.

[9] A. Mantri, S. Nandi, G. Kumar, and S.Kumar, "High Performance Architecture and Grid Computing: Springer Heidelberg Dordrecht London NewYork", 2011.

[10] M. Nazir, P. Tiwari, S. D. Tiwari, and R. G. Mishra, "Cloud Computing: An Overview",*Book Chapter of Cloud Computing Reviews, Surveys, Tools, Techniques and Applications: An Open-Access eBook published by HCTL Open*, 2015.

7

Capacity Planning in Cloud

Learning Objectives

After reading this chapter, you will be able to

- Illustrate the need for capacity planning in Cloud

- Analyze the elements involved in the capacity planning of Cloud infrastructures

- Challenges of capacity planning in Cloud computing environment

Cloud computing offers incredible flexibility and fantastic options compared to traditional computing or storage methods. Many Cloud services operate with pay-as-you-go models that have no boundaries on capacity or usage. The question arises: Is capacity planning necessary in the Cloud? One of the biggest advantages Cloud computing offers is the elastic usage of resources. The Cloud requires less hardware than traditional computing structures which means greater flexibility and a lower upfront cost. Even better, it can be easy and quick to purchase additional resources. However, elasticity is only good if our implementation strategy is good.

Even though increased capacity can be easily acquired, it is important to know what our organization is currently using and will need in the future. With the Cloud, the possibility of elasticity is ever present but it is up to the user to know how to use that power effectively. Capacity planning measures the maximum amount of work that can be done using the current technology (Fig. 7.1) and then adds resources to do more work as needed in the Data Centers (DCs). Figure 7.2 shows the self-explanatory steps involved in capacity planning. It conveys that the capacity planning measures the maximum amount of work that can be done using the current technology and then adds resources to do more work as needed.

Preliminaries

The following concepts discussed here provides easier understanding in the remaining part of this chapter.

CAPEX vs OPEX: The traditional mode of capacity planning focuses on obtaining servers funded by applications able to achieve capital investment.

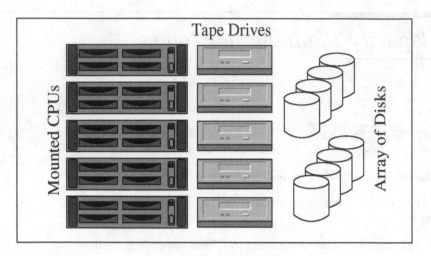

FIGURE 7.1
Cloud Data Center

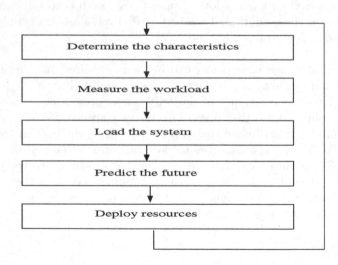

FIGURE 7.2
Capacity planning

Application groups had to obtain the capital needed to fund compute resources to operate the application. Capital expenditures (CAPEX), operational expenditures (OPEX) simplified is like choosing between purchasing a car on full payment with yearly depreciation benefits, or leasing a vehicle on a monthly cost with some limits on miles.

Capacity planning: It looks to coordinate demands to accessible assets. Capacity planning looks at what frameworks are set up, measures their exhibition, and decides designs in use that empowers the organizer to anticipate request. Assets are provisioned and apportioned to satisfy need.

Power consumption: The costs of labor, hardware, facilities, and network are relatively stable for a short term. These components are not suitable for dynamic adjustment at a fine granularity since the adjustment overhead is even more than the amount of cost savings. But the power supply can be adjusted readily. It is necessary to control the power consumption to reduce the cost of the Cloud as well as reduce the carbon dioxide emission. The US Environment Protection Agency (EPA) estimates that the energy usage at Cloud DCs is successively doubling every five years. Based on current trend estimates, U.S. DCs are projected to consume approximately 73 billion kWh in 2020.

7.1   Cloud Capacity Model

The traditional capacity-planning process is typically achieved in four steps. They are: (i) Creating a capacity model that defines the key resources and units of growth, (ii) Creating a baseline to understand how the server, storage, and network infrastructure are used by capturing secondary indicators such as CPU load or global network traffic, (iii) Evaluating changes from new applications that are going to run on the infrastructure and the impact of demand "bursting" because of increased activity in given services, and (iv) Analyzing the data from the previous steps to forecast future infrastructure requirements and decide how to satisfy these requirements. The main challenge with this approach is that it is very focused on the technology "silos," not the platform as a whole.

Companies are consistently being challenged today to reduce their overall expense on technology. Cloud computing is one of the key methods of transformation that organizations are using to accomplish their goal. The Cloud enables companies to function on an operating expense (OPEX) budget rather than a capital expense (CAPEX) budget. Cloud computing provides faster time to market, chargeback ability, and the capability to spin up and spin down virtual machines (VMs) to meet fluctuating requirements. The Cloud is an infinitely-scalable environment, bursting with potential. However, speed, scalability, and virtualization do not guarantee a Return-on-investment (ROI). All the benefits of the Cloud remain potential instead of actual, if the Cloud is not managed for maximum efficiency and cost-effectiveness. This brings companies who are in the Cloud or who are considering a move to the Cloud to a very important point discussed as follows.

Service Level Agreements

The challenge when analyzing SLAs is that application owners, business functions, and IT may have very different expectations based upon considerations such as perceived criticality, complexity and cost. Working independently from one another, these various areas could assume contradictory SLAs or prioritization levels for the same item, resulting in misunderstanding, conflict, and potentially increased business risk. It is imperative for all stakeholders to achieve consensus for each SLA that is assigned, based upon a thorough review of the needs and involved factors.

Utilization patterns

In the financial world, certain applications will be running at full capacity during the market's hours of operation. But the moment the market closes, the utilization for those applications is decreased. IT should therefore, seek to use the same infrastructure to run resource-intensive applications, perhaps batch jobs or fixed income applications. Doubling up in this strategic fashion removes the wastage (and cost) of two systems sitting idle 50% of the time by allowing one system to function at optimum capacity 100% of the time.

Workload analytics

Whereas utilization patterns reflect the current state of affairs, workload analytics explore the potential impact of an action on the existing environment, before that action takes place. For instance, businesses might create workload scenarios to identify the ramifications of bringing on a new product, service, or technology. They might forecast the effects of penetrating a new market. A company that is highly seasonal could estimate the difference between utilization in slow months versus heavy volume months. These scenarios help determine trending and define the additional capacity that will be required to handle increased workloads.

Exceptions process

Not everything can or should be moved to the Cloud. Companies should therefore establish a well-defined exceptions process. An exceptions process would ask questions, such as: (i) Is this software supported on a virtualized platform?, (ii) Is the necessary agility and scalability assured in the proposed Cloud environment, or are there potential limitations that could affect the performance of a given application?, (iii) Is there a memory footprint that this application supports that cannot be handled in a Cloud computing environment?, (iv) Will security, compliance, and regulatory requirements be satisfied if this application and its associated data are moved to the Cloud?, (v) Is the application integrated with other applications, and if so, will the Cloud support such integration?, (vi) What availability is required for this application, and will the Cloud support the necessary SLAs?

Data management

Before data is moved to the Cloud, policies addressing the creation, access, retention, archiving, and deletion of the data should be set in place to establish and maintain an optimized environment. Data may need to be re-tiered, moved to low-cost storage, or deleted entirely from the system on a regular basis. Without data management policies, it is easy to end up with multiple files containing the same data, and to keep all data forever.

Configuration

The Cloud, like almost anything else, can be customized. But customization always costs. Before considering customization, businesses should examine the standard configurations available with their intended Cloud environment and determine whether these pre-defined configurations will be appropriate for the applications and data being migrated to the Cloud. If they are, that will save time, effort, complexity, and expense in both Cloud migration and Cloud maintenance.

Business requirements

One factor, namely, IT always has the responsibility to convert business requirements into technology solutions. For example, the sales department may indicate that they need to process 30 transactions per minute. IT must take that business requirement and break down the impact on the compute cycle, the memory footprint, the input/output (IO), the drives, etc. Based upon that detailed information, a technology solution can be designed and implemented.

To engage in effective Cloud computing capacity planning, IT needs to sit at the table with the rest of the business, hear their needs in their language, translate those needs into IT terminology, and deliver for the business. IT also needs to reverse that communication process many times; for example, they may need to explain a technology-oriented disaster recovery policy to the rest of the business in terms the recipients can understand and value. Only when the business is working together as a unified whole can the Cloud environment be truly optimized.

Capacity management in Clouds

On one hand, underestimation of the Cloud capacity can cause resource shortage and revenue loss. On the other hand, overestimation of the Cloud capacity may result in idled resources and unnecessary utility costs. Unused hardware not only causes under-utilized capital, but also results in more early purchase costs as the price of the same computing equipment is always going down. Figures 7.3 and 7.4 shows capacity versus usage comparison in traditional DC and Amazon web services, respectively. It is clear that the customer's dissatisfaction has reduced in the recent years, as the requirement of resources are nearly met at all times. This is attributed to the efficient capacity planning in the Cloud servers.

Resource allocation

A key issue in the administration of Cloud DC assets is to recognize assets that can meet the necessities of each service and allot those assets to the services in a manner that is good with the Cloud supplier's objectives. This is accomplished by an allocation procedure whose quality is frequently estimated utilizing an utility capacity that represents Cloud provider's objectives. The ideal asset allotment mechanism is what amplifies this utility capacity. The utility capacity might be characterized to underscore benefit, asset use, vitality proficiency, or the reasonableness of the allotment plot. Allotment procedures ought to be dynamic since application's remaining tasks at hand shift powerfully after some time and their lifetimes are commonly obscure when assets are at first dispensed. Because of its dynamic nature, the assignment procedure may profit by re-consolidation and over-consolidation of assets.

Elasticity

The essential component that recognizes Clouds from other conditions is their ability for on-request service provisioning, which is known as elasticity. Cloud clients ought to have the option to provision and de-provision their resources as their demand changes, with the end goal that the accessible assets match the current interest as conceivable at each point in time. Flexibility calculations depend on checking information to evaluate request dependent on the current load and can be partitioned in two sorts: those that change the quantity of VM cases of a specific kind and along these lines provide horizontal elasticity, and those that change the size of the VMs and in this manner give vertical elasticity.

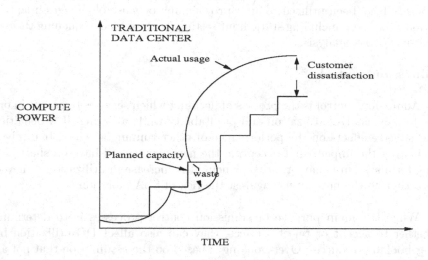

FIGURE 7.3

Capacity vs Usage comparison in Traditional DC

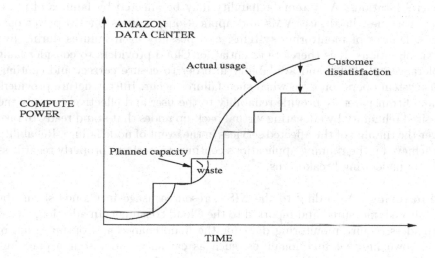

FIGURE 7.4

Capacity vs Usage comparison in Amazon web services

Various creators have contemplated versatility and a wide scope of approaches have been utilized to alter provisioning because of changes in application's requests, including static limit setting, control theory, queuing theory, and time series analysis.

Admission control

Admission control is the process of deciding which services to accept in order to increase DC utilization and profitability while avoiding SLA penalties and adverse effects on the performance of other running services. It involves evaluating the impact that accepting the new service will have on short and long-term system behavior, i.e., balancing the increase in utilization achieved by accepting the new service against the risk of SLA violations.

While the main purpose of admission control processes is to determine whether to accept or reject services, they can also affect DC utilization by over-booking resources. Over-booking is based on the assumption that not all of the resources requested by services will be completely utilized during those service's lifetimes.

Reliability and Fault Management

Due to their great complexity, even very carefully engineered DCs experience a large number of failures, especially when they are distributed over several locations. A system's reliability may be affected by failures of physical machines, hardware, VMs and applications, failures due to power outages, failures of monitoring systems, and possibly even failures during live VM migrations. It is therefore essential for Cloud providers to consider fault tolerance, reliability and availability in order to ensure correct and continuous system operation even when such failures occur. Infrastructure providers should transparently provide reliability to the user and effectively supply the desired reliability by allocating virtual back-up nodes that stand ready to take over the running of the affected services in the event of node failure. Reliability is achieved if the running applications continue to perform properly regardless of any underlying breakdowns.

Monitoring   According to the NIST, resource usage in Clouds should be monitored, measured and reported to the Cloud consumer. In addition, Cloud providers require monitoring data for the maintenance and optimization of their own internal functionalities, such as capacity and resource planning, SLA management, billing, trouble-shooting, performance management, and security management. Cloud monitoring can be performed on different levels, each of which requires its own metrics and provides specific information. For example, hardware level monitoring may provide information on factors such as CPU and memory utilization while application level monitoring may provide information on an application's response times or throughput.

7.2 Probabilistic Auto-scaling for Capacity Planning

In this section, we describe the auto-scaling approach for capacity planning. The main idea of this approach is that each autonomic service decides to create new autonomic services or remove itself in a probabilistic manner and independent of other autonomic services. The purpose is to have a total number of autonomic services such that the utilization of each autonomic service stays close to a given threshold. The auto-scaling algorithm is shown in Algorithm 7.1. In charge of executing it is the *Autonomic Manager* subcomponent of the autonomic service given by Cloud service provider. The Autonomic Manager periodically (with period equal to T^s) retrieves the neighborhood load (averaged over the last T^m time frame). If the load is less than the minimum load threshold (L^{min}), then the possibility to remove the current autonomic service is considered in Algorithm 7.2. Otherwise, if the average neighborhood load is higher than the maximum load threshold (L^{max}) then the autonomic service tries to add new autonomic services (see Algorithm 7.3).

Algorithm 1 Algorithm 7.1: Probabilistic Algorithm for Scaling

Description:
> **while (true) do**
> $wait(T^s)$
> $L \leftarrow monitor.computeAverageNeighborhoodLoad(T^m)$
>   **if** $L < L^{min}$
> $analyzeRemoval(L)$
>   **else**
>   **if** $L > L^{max}$
> $analyzeAddition(L)$
>   **end if**
>   **end if**
>   **end while**

Both *analyzeRemoval* and *analyzeAddition* functions rely on the *computeRatio()* function to be computed from equation 7.5. This formula is highly parameterized. As this ratio is negative, the *analyzeRemoval* function computes its absolute value which is used as a probability for removing the current autonomic service. In the *analyzeAddition* function, this ratio can be a value where the integer part represents the number of autonomic services to be added while the fractional part is used as the probability for adding an extra autonomic service.

To derive equation 7.5, a set of parameters required are summarized in Table 7.1. Two important parameters of an autonomic service are the capacity and the load. The capacity (denoted with C_i) is the maximum number of requests (jobs) per second that can be processed by the autonomic service

---

Algorithm 2 Algorithm 2: analyzeRemoval(L)

Description:

$r \leftarrow computeRatio(L)$
  if $abs(r) > random()$
  $removeSelf()$
  end if

Algorithm 3 Algorithm 3: analyzeAddition(L)

Description:

$r \leftarrow computeRatio(L)$
$n \leftarrow \lfloor r \rfloor$
$\{r\} < random()$
$n \leftarrow n + 1$
  end if

---

i and is derived through benchmarking. The load of an autonomic service i (denoted with $L_i(t)$) computed at time t is the ratio between the average number of requests per second that were issued in the time interval $[t - T^m, t]$ and the capacity of the autonomic service. All the load parameters, such as the average load per autonomic service of the system and the load thresholds are also expressed as percentages with respect to the capacity of the autonomic service.

Given at time t the actual number of nodes of the system $N(t)$, the average load $L^{av}(t)$, a desired target load L^{des}, and an average capacity C^{av} that is always constant independently from the number of the nodes, then the target number of nodes at time t is calculated as given in Equation 7.1.

$$\hat{N}(t) = \frac{L^{av}(t)}{L^{des}} \cdot N(t) \tag{7.1}$$

Equation 7.1 requires redistribution of the total load of the system (defined as $L^{av}(t) \cdot N(t)$) in a system in which the autonomic services have different weights in terms of load (from $L^{av}(t)$ to L^{des}). Therefore the total number of autonomic services to be added (positive value) or removed (negative value) from the system is denoted by Equation 7.2.

$$M(t) = \hat{N}(t) - N(t) = \frac{L^{av}(t) - L^{des}}{L^{des}} \cdot N(t) \tag{7.2}$$

Computing $M(t)$ in presence of global information such as $N(t)$ and $L^{av}(t)$ is very simple, but this system aim to provide a decentralized solution in which each autonomic service decides to add new autonomic services or remove itself in a probabilistic way. The ratio $R_i(t) = M(t)/N(t)$ can be used for deriving this probability. By making these simplifications equation 7.3 is obtained:

TABLE 7.1

Auto-scaling notations

$N(t)$	Actual number of autonomic services allocated at a given time
$\hat{N}(t)$	Target number of autonomic services to be allocated at a given time
$M(t)$	Number of autonomic services to be added/removed at a given time
C_i	Capacity of autonomic service i
C^{av}	Average capacity among all system autonomic services
T^m	Length of monitoring timeframe for the actual load
$L_i(t)$	Load of autonomic service i computed at time t over timeframe $(t-T^m, t)$ (percent with respect to autonomic service capacity)
$L^{av}(t)$	Average load per autonomic service of the system, computed at time t over timeframe $(t-T^m, t)$ for all autonomic services in the system (percent)
$\tilde{L}_i^{av}(t)$	Average load per autonomic service of the neighborhood of autonomic service i, computed at time t over timeframe $(t-T^m, t)$ (percent)
L^{min}	Minimum load threshold (percent)
L^{max}	Maximum load threshold (percent)
L^{des}	Desired load threshold, which is equal to $(L^{max} - L^{min})/2$ (percent)
T^s	Period between two successive runnings of the auto-scaling algorithm on a autonomic service
$neig_i$	Neighborhood of autonomic service i: contains autonomic service i and its neighbors
$queue_i$	Number of enqueued requests in autonomic service i
R_{max}	Maximum response time for completed requests (from SLA)

$$R_i(t) = \frac{M(t)}{N(t)} = \frac{L^{av}(t) - L^{des}}{L^{des}} \tag{7.3}$$

During the simulation process, each autonomic service uses the weighted average load of its neighborhood as an approximation of the average load of the system using capacities as weights (see equation 7.4).

$$\tilde{L}_i^{av}(t) = \frac{\sum_{j \in neig_i}(C_j \cdot L_j(t))}{\sum_{i \in neig_i} C_j} \tag{7.4}$$

Then, this value can be used to calculate a local estimator for the ratio $R_i(t_1)$, that we have denoted in equation 7.5 as $\tilde{R}_i(t_1)$ and used in algorithms 7.2 and 7.3, as the *ComputeRatio()* function.

$$\tilde{R}_i(t) = \frac{\tilde{L}_i^{av}(t) - L^{des}}{L^{des}} \tag{7.5}$$

When $\tilde{R}_i(t) < 0$, autonomic service i uses $|\tilde{R}_i(t)|$ as a probability to remove itself. When $\tilde{R}_i(t) > 0$, the fractional part of $\tilde{R}_i(t)$ is used as a probability to add a new autonomic service, while the integer part (which is greater than zero when the number of autonomic services to be added is higher than the number of existing autonomic services) is the number of autonomic services that will be added in a deterministic way. The capacity of the new autonomic service is assumed to be, in the average case, approximated to the average capacity in the neighborhood of the local autonomic service.

In addition to the decision rules for adding and removing resources to the system, the probabilistic auto-scaling approach requires the adoption of proper fault-tolerant mechanisms to keep the load among the autonomic services balanced and the topology connected.

Energy efficiency in Clouds: Cloud computing is a fundamentally more efficient way to operate compute infrastructure. The increase in efficiency driven by the Cloud are due to increased utilization. All companies have to provision their compute infrastructure for peak usage. But, they only monetize the actual usage which goes up and down over time. What this leads to incredibly low average utilization levels with 30% being extraordinarily high and 10 to 20% the norm. Cloud computing gets an easy win on this dimension. When non-correlated workloads from a diverse set of customers are hosted in the Cloud, the peak to average flattens dramatically. Immediately effective utilization sky rockets. Where 70 to 80% of the resources are usually wasted, this number climbs rapidly with scale in Cloud computing services flattening the peak to average ratio.

7.3   Advantages of Capacity Planning

Comprehensive Cloud computing capacity planning brings an organization immediate value, ongoing Return-on-Investment (ROI), and the opportunity for continuous improvement. They are discussed as follows.

Prioritization: In the short term, Cloud computing capacity planning enables strategic prioritization. Businesses can assess the true ROI with regard to eliminating hardware that is unusable, replacing software that is no longer being supported by the vendor, combining applications in a shared environment, retiring applications that are obsolete, refreshing technology on an opex basis, and leveraging new technology, software, or applications.

Right-sizing: Despite the fact that the essence of the Cloud gives more profit with less investments, companies can fall into the trap of assuming that they need the same capacity in the Cloud as they do in a traditional DC to run their operations. Cloud computing capacity planning removes the potential for this error and empowers a company to right-size its IT environment by setting down in quantifiable form, which is needed in the Cloud, based upon the company's unique requirements. For example, suppose a department currently runs an application on a standard physical server with four core CPUs and 2 GB of memory. Cloud computing capacity planning might show that the application is only using 10% of this environment. Therefore, in the new Cloud environment, one virtual core CPU with 500 MB of memory might be sufficient for the few times the application actually runs.

Virtualization 2.0: One of the biggest issues encountered when companies move to the Cloud is that they create an infrastructure optimization plan to reduce cost, migrate to the Cloud, achieve some level of virtualization and then stop. Migration is a project. But if a company continues to leverage Cloud computing capacity planning on a go-forward basis, the regular analysis will identify additional opportunities for cost savings, efficiencies, productivity and agility as they open up. Businesses can move toward Virtualization 2.0.

Measuring and managing ROI: In business, we cannot manage if we cannot measure. Cloud computing capacity planning allows organizations to do both by delineating a framework for gathering and analyzing key data. It enables decision-makers to answer the questions of what, where, when, why and how to leverage the Cloud to meet their business requirements. ROI is then a matter of acting on that data-both in the initial migration and on a consistent basis going forward.

Summary

Capacity planning is a significant action that guarantees clients to have the required processing assets available whenever required. The present high performing data centers have a lot of memory and storage capacities that can best be completely utilized through virtualization techniques. This asset rich IT condition has prompted new and better approaches to get ready for better approaches of required processing requirements for the present business applications.

Keywords

Capacity planning	Resource utilization
Prediction	Management

Objective type questions

1. Point out the correct statement :

(a) Capacity planning seeks to match demand to available resources

(b) Capacity planning examines what systems are in place, measures their performance, and determines patterns in usage that enables the planner to predict demand

(c) Resources are provisioned and allocated to meet demand

(d) All of the above.

2. What is the relation of capacity attribute to performance ?

(a) same

(b) different

(c) ambiguous

(d) None of the above.

3. Geographic distribution of data across a Cloud provider's network is a problem for many enterprises because it

(a) Breaks compliance regulations

(b) Adds latency

(c) Raises security concerns

(d) Makes data recovery harder

4. Amazon Web Services is which type of Cloud computing distribution model?

(a) Software as a Service

(b) Platform as a Service

(c) Infrastructure as a Service

(d) None of the above.

5. Which of the following is critical step in capacity planning ?

(a) Predict the future based on historical trends and other factors

(b) Load the system until it is overloaded

(c) Deploy or tear down resources to meet your predictions

(d) All of the above.

6. A good Cloud computing network can be adjusted to provide bandwidth on demand

(a) True

(b) False

7. A larger Cloud network can be built as either a layer 3 or layer 4 network

(a) True

(b) False

8. The typical three-layer switching topology will not create latency within a Cloud network

(a) a. True   (b) b. False

9. Which of the following is hardest factor to determine ?

(a) Network performance   (b) Network capacity

(c) Network delay   (d) All of the above.

10. Point out the wrong statement:

(a) Network capacity is one of the easiest factors to determine

(b) A system uses resources to satisfy cloud computing demands that include processor, memory, storage, and network capacity

(c) Scaling a system can be done by scaling up vertically to more powerful systems

(d) All of the above.

11. Which of the following is used as Open source distributed monitoring system ?

(a) Dstat   (b) GraphClick

(c) Ganglia   (d) None of the above.

12. Which of the following command is used to display the level of CPU activity in Linux ?

(a) sep   (b) sav

(c) sar   (d) None of the above.

13. Which of the following tool captures time-dependent performance data from resources such as a CPU load ?

(a) RBTool   (b) RRDTool

(c) Perfmon   (d) None of the above.

14. Point out the wrong statement.

(a) The total workload might be served by a single server instance in the Cloud

(b) Performance logs are the only source of performance measurements

(c) The amount of resources to be deployed depends upon the characterization of the Web servers involved, their potential utilization rates, and other factors

(d) All of the above.

15. Which of the following tool is used for Web site monitoring service in LAMP ?

(a) Alertra

(b) Cacti

(c) Collectd

(d) None of the above.

Objective type questions -answer
1:d 2:b 3:a 4:c 5:b 6:a 7:a 8:b 9:b 10:a 11:c 12:c 13:b 14:b 15:a

Review questions

1. How buffer is used in Amazon web services?
2. In Cloud computing environment, what are the optimizing strategies?
3. List down the basic characteristics of Cloud computing.
4. What is the use of API's in Cloud services?
5. Mention the name of some large Cloud providers and databases.
6. Which of the tool captures time-dependent performance data from resources such as a CPU load?
7. Which of the testing seeks to answer maximum load that my current system can support?
8. What is the unused capacity in virtual infrastructures?

Critical thinking questions

1. Discuss the implications of the exponential improvement of computer and communication technologies on system complexity.
2. Discuss means to cope with the complexity of computer and communication systems other than modularity, layering and hierarchy.
3. Analyze the reasons for the introduction of storage area networks (SANs) and their properties.
4. There are platform-as-a-service plays for popular application stacks such as Java (VMForce.com and Google App Engine), .Net (Microsoft Azure) and Ruby-on-Rails (Heroku and Engine Yard), so why not the LAMP stack?

Bibliography

[1] Bruno Abrahao, Virgilio Almeida, Jussara Almeida, Alex Zhang, Dirk Beyer, and Fereydoon Safai, "Self-adaptive sla-driven capacity management for Internet services", *Proceedings of Network Operations and Management Symposium, 2006. NOMS 2006. 10th IEEE/IFIP*, pages 557-568, April 2006.

[2] Arshdeep Bahga and Vijay Krishna Madisetti, "Synthetic workload generation for Cloud computing applications", *Journal of Software Engineering and Applications*, 4(7):396-410, July 2011.

[3] Ludmila Cherkasova, Wenting Tang, and Sharad Singhal, "An SLA-oriented capacity planning tool for streaming media services", *Proceedings of 2004 International Conference on Dependable Systems and Networks*, pages 743-752, Washington, DC, USA, 2004. IEEE Computer Society.

[4] Tony Clark, Andy Evans, Stuart Kent, and Paul Sammut, "The MMF approach to engineering object-oriented design languages", *Proceedings of the Workshop on Language Descriptions*, Tools and Applications, April 2001.

[5] Christina Delimitrou and Christos Kozyrakis, "Cross-examination of datacenter workload modeling techniques", *Proceedings of 31st International Conference on Distributed Computing Systems Workshops (ICDCSW)*, pages 72 -79, June 2011.

[6] Brian Dougherty, Jules White, and Douglas C. Schmidt, "Model-driven auto-scaling of green Cloud computing infrastructure", *Future Generation Computer Systems*, 28(2):371-378, 2012.

[7] Jorge Ejarque, Marc de Palol, Inigo Goiri, Ferran Julia, Jordi Guitart, Rosa M. Badia, and Jordi Torres, "SLA-driven semantically-enhanced dynamic resource allocator for virtualized service providers", *Proceedings of Fourth IEEE International Conference on eScience*, 2008.

[8] Archana Ganapathi, Yanpei Chen, Armando Fox, Randy Katz, and David Patterson, "Statistics-driven workload modeling for the Cloud", *Procdings of IEEE 26th International Conference on Data Engineering Workshops (ICDEW)*, pages 87-92. IEEE, March 2010.

[9] Steven Kelly and Juha-Pekka Tolvanen, "Domain-Specific Modeling: Enabling Full Code Generation", *Wiley-IEEE Computer Society*, March 2008.

[10] Hamzeh Khazaei, Jelena Misic, and Vojislav B. Misic, "Modelling of Cloud computing centers using M/G/M queues", *Proceedings of 31st International Conference on Distributed Computing Systems Workshops (ICDCSW)*, pages 87-92, June 2011.

[11] Xue Liu, Xiaoyun Zhu, Sharad Singhal, and Martin Arlitt, "Adaptive entitlement control of resource containers on shared servers", *Proceedings of 9th IFIP/IEEE International Symposium on Integrated Network Management*, 2005.

[12] C.C.T. Mark, D. Niyato, and Tham Chen-Khong, "Evolutionary optimal virtual machine placement and demand forecaster for Cloud computing", *Proceedings of IEEE International Conference on Advanced Information Networking and Applications (AINA)*, pages 348-355, March 2011.

8

SLA Management in Cloud Computing

Learning Objectives

After reading this chapter, you will be able to

- Summarize the need and characteristics of Service-Level-Agreement (SLA)

- Describe SLA monitoring in Cloud

- Identify research opportunities in SLA management

The Service-Level-Agreement (SLA) is an agreement between two or more parties, where one is the customer and the others are service providers. This can be a legally binding formal or an informal "contract" (for example, internal department relationships). The agreement may involve separate organizations, or different teams within one organization. Contracts between the service provider and other third parties are often (incorrectly) called SLAs, because the level of service has been set by the (principal) customer, there can be no "agreement" between third parties; these agreements are simply "contracts". Operational-level agreements (OLAs), however, may be used by internal groups to support SLAs. If some aspect of a service has not been agreed with the customer, it is not an "SLA".

SLAs are, by their nature, "output" based, the result of the service as received by the customer is the subject of the "agreement." The (expert) service provider can demonstrate their value by organizing themselves with ingenuity, capability, and knowledge to deliver the service required, perhaps in an innovative way. Organizations can also specify the way the service is to be delivered, through a specification (a service level specification) and using subordinate "objectives" other than those related to the level of service. This type of agreement is known as an "input" SLA. This latter type of requirement is becoming obsolete as organizations become more demanding and shift the delivery methodology risk on to the service provider.

Preliminaries

The following are some of concepts discussed here for easier understanding in the remaining part of this chapter.

SLA basics The following basics should be borne in mind for better understanding of SLAs. The fulfillment of an service level objectives (SLOs) describes a state of service when all of the SLOs key performance indicators

143

are within a specified thresholds. Key public infrastructures (KPIs) usually consist of one or more raw monitored values including min., avg. and max. specifying the scale. They can also represent some aggregated measurement (e.g., average output) within a sliding window that is combined from one or more monitoring outputs. The Cloud computing infrastructures are usually large scale, therefore SLAs need to be formally described to enable their automated handling and protection.

SLA requirements Following are few of the aspects that the standard SLA aims to satisfy: (i) A signed agreement with each customer, (ii) Transactions by hour and jobs by day for each application, (iii) A method of reporting SLA results, (iv) Priority of services in case of insufficient availability, (v) Agreed methods of penalty in case customer exceeds his limits, (vi) Agreed methods of penalty in case Cloud provider fails to meet contract specifications and (vii) Schedule of virtual or actual meeting between the customer and the Cloud provider, if necessary.

Service Levels Service level agreements are defined at different levels.

(i) Customer-based SLA: An agreement with an individual customer group, covering all the services they use.

(ii) Service-based SLA: An agreement for all customers using the services being delivered by the service provider.

(iii) Multilevel SLA: The SLA is split into the different levels, each addressing different set of customers for the same services, in the same SLA.

8.1 Components of SLA

A well defined and typical SLA will contain the following components.

Type of service to be provided: It specifies the type of service and any additional details of type of service to be provided. In case of an IP network connectivity, type of service will describe functions such as operation and maintenance of networking equipments, connection bandwidth to be provided, etc.

The service's desired performance level, especially its reliability and responsiveness: A reliable service will be the one which suffers minimum disruptions in a specific amount of time and is available at almost all times. A service with good responsiveness will perform the desired action promptly after the customer requests for it.

Monitoring process and service level reporting: This component describes how the performance levels are supervised and monitored. This process involves gathering of different type of statistics, how frequently this statistics will be collected and how this statistics will be accessed by the customers.

The steps for reporting issues with the service: This component will specify the contact details to report the problem to and the order in which details about the issue have to be reported. The contract will also include a time range in which the problem will be looked upon and also till when the issue will be resolved.

Response and issue resolution time-frame: Response time-frame is the time period by which the service provider will start the investigation of the issue. Issue resolution time-frame is the time period by which the current service issue will be resolved and fixed.

Repercussions for service provider not meeting its commitment: If the provider is not able to meet the requirements as stated in SLA then service provider will have to face consequences for the same. These consequences may include customer's right to terminate the contract or ask for a refund for losses incurred by the customer due to failure of service.

8.1.1 Common Metrics in SLA

A common service in IT service management is a call center or service desk. Performance metrics commonly agreed between customers and service providers in these cases include the following:

- Abandonment Rate: It is the percentage of calls abandoned while waiting to be answered.

- Average Speed to Answer (ASA): Average time (usually in seconds) it takes for a call to be answered by the service desk.

- Time Service Factor (TSF): Percentage of calls answered within a definite timeframe, e.g., 80% in 20 seconds.

- First-Call Resolution (FCR): Percentage of incoming calls that can be resolved without the use of a callback or without having the caller call back the helpdesk to finish resolving the case.

Non–Compliance

Guarntees   Warranties   Contract
Serviceability
Penalties   **SLA**   Operation
Responsibilities   Quality   Goals
Availability   Performance   Scope

SERVICE LEVEL   Objectives
AGREEMENT   Innovative

Performance measurement

Communications

FIGURE 8.1
SLA metrics

- Turn-Around Time (TAT): Time taken to complete a certain task.

- Mean Time To Recover (MTTR): Time taken to recover after an outage of service.

Uptime is also a common metric, often used for data services such as shared hosting, virtual private servers and dedicated servers. Common agreements include percentage of network uptime, power uptime, number of scheduled maintenance windows, etc.

In case of Cloud computing environment, SLAs can contain numerous service performance metrics (Figure 8.1) with corresponding SLOs. When the Government accountability office (in US) issued a report calling for use of SLA for Cloud computing contracts, it identified key practices to be included in an SLA, which, if properly implemented, could help agencies "ensure services are performed effectively, efficiently and securely."

According to recently serached for information, the working group identified ten areas where SLA metrics are required: accessibility, availability, performance, service reliability, data management, attestations, certs and audits, change management, Cloud service support, governance and termination of service. Many SLAs track to the Information Technology Infrastructure Library specifications when applied to IT services. The set of SLA checklist is shown in Figure 8.2.

SLA CHECKLIST

> 1. Statement of Objectives
>
> 2. Scope of services to be covered
>
> 3. Service provider responsibilities
>
> 4. Customer responsibilities
>
> 5. Performance metrics (response time, resolution time, etc.)
>
> 6. Penalities for contract breach/exclusions.

FIGURE 8.2
SLA checklist

8.1.2   Specific Example

Few examples of SLAs tie-up for existing technologies are as follows.

Backbone Internet providers: It is not uncommon for an Internet backbone service provider (or network service provider) to explicitly state its own SLA on its website. The U.S. Telecommunications Act of 1996 does not expressly mandate that companies have SLAs, but it does provide a framework for firms to do so in Sections 251 and 252. Section 252(c)(1) for example ("Duty to Negotiate") requires Incumbent local exchange carriers (ILECs) to negotiate in good faith about matters such as resale and access to rights of way.

Web-service level agreement (WSLA): It is a standard for service level agreement compliance monitoring of web services. It allows authors to specify the performance metrics associated with a web service application, desired performance targets, and actions that should be performed when performance is not met.

Cloud computing: The underlying benefit of Cloud computing is shared resources, which is supported by the underlying nature of a shared infrastructure environment. Thus, SLAs span across the Cloud and are offered by service providers as a service-based agreement rather than a customer-based agreement. Measuring, monitoring and reporting on Cloud performance is based on the end Users experience (UX) or their ability to consume resources. The downside of Cloud computing relative to SLAs is the difficulty in determining the root cause of service interruptions due to the complex nature of the environment.

As applications are moved from dedicated hardware into the Cloud, they need to achieve the same or even more demanding levels of service than classical installations. SLAs for Cloud services focus on characteristics of the data center and more recently include characteristics of the network to support end-to-end SLAs. Any SLA management strategy considers two well-differentiated phases: negotiating the contract and monitoring its fulfilment in real time. Thus, SLA management encompasses the SLA contract definition: the basic schema with the Quality-of-Service (QoS) parameters; SLA negotiation; SLA monitoring; SLA violation detection; and SLA enforcement according to defined policies.

Outsourcing: It involves the transfer of responsibility from an organization to a supplier. This new arrangement is managed through a contract that may include one or more SLAs. The contract may involve financial penalties and the right to terminate if any of the SLAs metrics are consistently missed. Setting, tracking and managing SLAs is an important part of the outsourcing relationship management (ORM) discipline. Specific SLAs are typically negotiated up front as part of the outsourcing contract and used as one of the primary tools of outsourcing governance.

In software development, specific SLAs can apply to application outsourcing contracts in line with standards in software quality, as well as recommendations provided by neutral organizations like Consortium for Information & Software Quality (CISQ), which has published numerous papers on the topic (such as Using Software Measurement in SLAs) that are available to the public.

There are two types of SLAs from the perspective of application hosting. These are described here.

Infrastructure SLA : The infrastructure provider manages and offers guarantees on availability of the infrastructure, namely, server machine, power, network connectivity, and so on. Enterprises manage themselves, their applications that are deployed on these server machines. The machines are leased to the customers and are isolated from machines of other customers.

Application SLA : In the application co-location hosting model, the server capacity is available to the applications based solely on their resource demands. Hence, the service providers are flexible in allocating and de-allocating computing resources among the co-located applications. Therefore, the service providers are also responsible for ensuring to meet their customer's application SLOs (Table 8.1).

It is also possible to have agreement over several SLAs for different scenarios, and swithover from one to the next during the course of implementation.

TABLE 8.1

Key Contractual elements of an Application SLA

Components	Description
Service level parameter metric	Web site response time (e.g., max of 3.5 sec per user request). Latency of Web Server (WS) (e.g., max of 0.2 sec per request) Latency of database (e.g., max of 0.5 sec per query)
Function	Average latency of Web site = (latency of web server 1 + latency of web server 2) /2 Web site response time = Average latency of web server + latency of database
Measurement directive	database latency
Service level objective	Service assurance

8.2 Life Cycle of SLA

Each SLA goes through a sequence of steps starting from identification of terms and conditions, activation and monitoring of the stated terms and conditions, and eventual termination of contract once the hosting relationship ceases to exist. Such a sequence of steps is called SLA life cycle and consists of the following five phases. They are: (i) Contract definition, (ii) Publishing and discovery, (iii) Negotiation, (iv) Operationalization, and (v) De-commissioning.

Here, we explain in detail each of these phases of SLA life cycle.

Contract Definition: Generally, service providers define a set of service offerings and corresponding SLAs using standard templates. These service offerings form a catalog. Individual SLAs for enterprises can be derived by customizing these base SLA templates.

Publication and Discovery: Service provider advertises these base service offerings through standard publication media, and the customers should be able to locate the service provider by searching the catalog. The customers can search different competitive offerings and shortlist a few that fulfill their requirements for further negotiation.

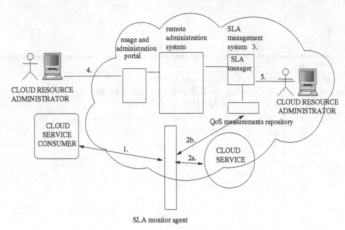

FIGURE 8.3
SLA management system

Negotiation: Once the customer has discovered a service provider who can meet their application hosting need, the SLA terms and conditions needs to be mutually agreed upon before signing the agreement for hosting the application. For a standard packaged application which is offered as service, this phase could be automated. For customized applications that are hosted on Cloud platforms, this phase is manual. The service provider needs to analyze the application's behavior with respect to scalability and performance before agreeing on the specification of SLA. At the end of this phase, the SLA is mutually agreed by both customer and provider and is eventually signed off. SLA negotiation can utilize the WS-negotiation specification.

Operationalization: SLA operation consists of SLA monitoring, SLA accounting, and SLA enforcement. SLA monitoring involves measuring parameter values and calculating the metrics defined as a part of SLA and determining the deviations. On identifying the deviations, the concerned parties are notified. SLA accounting involves capturing and archiving the SLA adherence for compliance. As part of accounting, the application's actual performance and the performance guaranteed as a part of SLA is reported. Apart from the frequency and the duration of the SLA breach, it should also provide the penalties paid for each SLA violation.

De-commissioning: SLA decommissioning involves termination of all activities performed under a particular SLA when the hosting relationship between the service provider and the service consumer has ended. SLA specifies the terms and conditions of contract termination and specifies situations under which the relationship between a service provider and a service consumer can be considered to be legally ended.

8.3 SLA Management System in Cloud

The SLA monitor mechanism is utilized to explicitly watch the runtime execution of Cloud administrations to guarantee that they are satisfying the legally binding QoS necessities distributed in SLAs (Figure 8.3). The information gathered by the SLA monitor is prepared by a SLA management system as a reporting metrices. This system can proactively fix or failover Cloud services when exception conditions occur, for example, when the SLA monitor reports a Cloud service as "down."

In Figure 8.3, the SLA monitor polls the Cloud service by sending over polling request messages. The monitor receives polling response messages that report that the service was "up" at each polling cycle and stores the "up" time-time period of all polling cycles 1 to N in the log database (Step 1). The SLA monitor polls the Cloud service that sends polling request messages. Polling response messages are not received (Step 2a). The response messages continue to time out, so the SLA monitor stores the "down" time-time period of all polling cycles N+1 to N+M in the log database (Step 2b). The SLA monitor sends a polling request message and receives the polling response message and stores the "up" time in the log database (Step 3). The Cloud resource administrator manages resources fulfillment for users (Steps 4 & 5).

Summary

Cloud computing has developed into a more acceptable computing paradigm for implementing scalable infrastructure resources given on-demand in a pay-by-use basis. Self-adaptable Cloud resources are needed to meet users application needs defined by Service Level Agreements (SLAs) and to limit the amount of human interactions with the processing environment. Sufficient SLA monitoring techniques and timely discovery of possible SLA violations are of principal importance for both Cloud providers and Cloud customers. In this chapter, identification of possible violations of SLA is discussed by analyzing predefined service level objectives together by using knowledgebases for managing and preventing such violations.

Keywords

SLA	SLA monitoring
SLA enforcement	Cloud Security

Objective type questions

1. Which of the following is specified parameter of SLA ?

(a) Response times   (b) Responsibilities of each party

(c) Warranties   (d) All of the above.

2. A _____ Level Agreement (SLA) is the contract for performance negotiated between you and a service provider.

(a) Service   (b) Application

(c) Deployment   (d) All of the above.

3. Which of the following provides tools for managing Windows servers and desktops ?

(a) Microsoft System Center   (b) System Service

(c) System Cloud   (d) All of the above.

4. Which of the following is not phase of Cloud life cycle management ?

(a) The definition of the service as a template for creating instances   (b) Client interactions with the service

(c) Management of the operation of instances and routine maintenance   (d) None of the above.

5. Point out the wrong statement:

(a) Google App Engine lets you deploy the application and monitor it   (b) From the standpoint of the client, a Cloud service provider is different than any other networked service

(c) The full range of network management capabilities may be brought to bear to solve mobile, desktop, and local server issues   (d) All of the above.

6. Which of the following is Virtual machine conversion Cloud ?

(a) BMC Cloud Computing Initiative   (b) Amazon CloudWatch

(c) AbiCloud   (d) None of the above.

7. _____ is Microsoft's Cloud-based management service for Windows systems.

(a) Intune

(b) Utunes

(c) Outtunes

(d) Windows Live Hotmail

8. Which of the following is a virtualization provisioning system that will be rolled into Dell's AIM ?

(a) Dell Scalent

(b) CloudKick

(c) Elastra

(d) All of the above.

9. Which of the following is used for Web site monitoring and analytics ?

(a) Gomez

(b) Ganglia

(c) Elastra

(d) None of the above.

10. Which of the following is used for performance management for virtualized Java Apps with VMware integration ?

(a) Hyperic

(b) Internetseer

(c) RightScale

(d) All of the above.

11. Which of the following is one of the property that differentiates Cloud computing ?

(a) scalability

(b) virtualization

(c) composability

(d) All of the above.

12. Point out the correct statement :

(a) Cloud computing is a natural extension of many of the design principles, protocols, plumbing and systems

(b) Platforms are used to create more easy software

(c) All SLAs are enforceable as contracts

(d) All of the above.

13. Which of the following language is used to manage transactions ?

(a) WSDL

(b) XML

(c) Soap

(d) All of the above.

14 Which of the architectural layer is used as backend in Cloud computing ?

(a) client

(b) Cloud

(c) soft

(d) All of the above.

15. A _____ Cloud requires virtualized storage to support the staging and storage of data.

(a) soft   (b) compute

(c) local   (d) None of the above.

Objective type questions -answer

   1:d   2:a   3:a   4:d   5:b   6:c   7:a   8:a   9:a   10:a   11:c   12:a   13:a   14:b   15:b

Review questions

1. Do customers publish SLAs, and how are these documents accessed?

2. How do our SLA targets differ from your competitors?

3. Why are SLA targets chosen? Targets are often defined competitively or based on the best or worst capability of the underlying products.

4. How often customers violate their SLAs with service providers in the span of 3 months/6 months/12 months?

5. Do customers publish their SLA results openly?

6. Which SLA metrics fail at most often, even if it has no impact on the customers?

7. How often do customers increase or decrease the SLA targets, and what has the trend been?

8. What SLA metrics have been removed in the last 12 months?

9. How often do customer test their own SLAs?

10. Do we use any third parties to monitor your SLAs? This can provide additional validation of the seriousness of SLA measurement?

Critical thinking questions

1. Evaluate the SLA toolkit at http://www.service-level-agreement.net/. Is the interactive guide useful, what does it miss? Does the SLA template include all clauses that are important in your view, what is missing? Are the examples helpful?

2. Peer-to-peer systems and Clouds share a few goals, but not the means to accomplish them. Compare the two classes of systems in terms of architecture, resource management, scope, and security.

3. Overprovisioning is the reliance on extra capacity to satisfy the needs of a large community of users when the average-to-peak resource demand ratio

is very high. Give an example of a large-scale system using overprovisioning and discuss if overprovisioning is sustainable in that case and what are the limitations of it. Is Cloud elasticity based on overprovisioning sustainable? Give the arguments to support your answer.

4. Software licensing is a major problem in Cloud computing. Discuss several ideas to prevent an administrator from hijacking the authorization to use a software licence.

Bibliography

[1] Mensce and V. Almeida, "Capacity Planning for web Performance: Metrics, Models and Methods", *Prentice-Hall, Englewood Cliffs*, NJ, 2018.

[2] P. Barham, B. Dragovic, K. Fraser, S. Hand, T. Harris, A. Ho, R. Neugebauer, I. Pratt, and A. Warfield, "Xen and the art of virtualization", *Proceedings of the 19th ACM Symposium on Operating Systems Principles*, New York, October 19À22, 2003, pp. 164-177.

[3] G. Popek and R. Goldberg, "Formal requirements for virtualizable third generation architectures", *Communications of ACM*, 17(7):412-421, 2014.

[4] E. de Souza E. Silva, and M. Gerla, "Load balancing in distributed systems with multiple classes and site constraints", *Proceedings of the 10th International symposium on Computer system modeling: Measurement and evaluation*, Paris, France, Dec. 2014, pp. 91-113.

[5] Verma, Dinesh, "Service level agreements on IP networks", *IEEE Communications on Networks*, 92 (9), 2017, pp. 215-218.

[6] "Global IP Network SLA", *NTT Communications*, Retrieved 22 June 2016.

[7] "Global Latency and Packet Delivery SLA", *Verizon web service*, Retrieved 22 June 2016.

[8] Rueda J.L, Gómez S.G, Chimento A.E, "The Service Aggregator Use Case Scenario", *Springer Science and Business Media*, LLC. pp. 329-342, 2017.

[9] Gallizo G, Kuebert R, Oberle K, Menychtas A, Konstanteli K, "Service Level Agreements in Virtualised Service Platforms", *Proceedings of the IIMC International Information Management Corporation*, pp. 1-8, 2018.

9

Resource Management in Cloud

Learning Objectives

After reading this chapter, you will be able to

- Elaborate the significant issues of resource management in Cloud

- Analyze the existing literatures w.r.t resource provisioning, resource allocation, resource adaptation and resource mapping in Cloud

- Understand the open challenges in resource management schemes.

Cloud computing is a new era Internet based computing where one can get to their own assets effectively from any PC through Internet. Cloud delivers computing as an utility, as it is accessible to the Cloud consumers on request. It is a basic pay-per-use model. It contains enormous number of shared assets. Resources are shared among multiple Cloud users, hence resource management is an issue in Cloud environment.

Due to the availability of finite resources, it is very challenging for Cloud providers to provide all the requested resources. From the Cloud providers perspective, Cloud resources must be allocated in a fair and efficient manner. In this chapter, we present significant research carried out in resource management for IaaS in Cloud computing and bring out the open challenges with respect to resource provisioning, resource allocation, resource adaptation and resource mapping.

Because of the availability of limited resources in data centers, it is challenging for Cloud sproviders to fulfill the requested resource. From the Cloud providers point of view, Cloud resources must be utilized in a reasonable and effective way. In this chapter, we present significant research carried out w.r.t resource provisioning, allocation, adaptation and mapping.

Preliminaries

The following are some of terms and terminologies given here for easier understanding in the remaining part of this chapter.

159

Resource provisioning: It is the allocation of a service provider's resources to a customer.

Resource allocation: It is the distribution of resources economically among competing groups of people or programs.

Resource adaptation: It is the ability or capacity of the system to adjust the resources dynamically to fulfill the requirements of the user.

Resource mapping: It is a correspondence between resources required by the users and resources available with the provider.

Resource modelling: It is a framework that illustrates the most important elements of resource management within a given environment.

Resource estimation: It is a close guess of the actual resources required for an application, usually done with some intuition or calculation.

Resource discovery and selection: It is the identification of list of authenticated resources that are available for job submission and to choose the best among them.

Resource brokering: It is the negotiation of the resources through an agent to ensure that the necessary resources are available at the right time to complete the objectives.

Resource scheduling: It determines when an activity should start or end, depending on its (1) duration, (2) predecessor activities, (3) predecessor relationships and (4) resources available.

9.1   Significant Resources in Cloud

In Cloud, service providers manage various resources. As Cloud computing is a utility based computing, this section throws light on the significant resources in Cloud.

Fast Computation Utility: This type of resource provides fast computational utility in Cloud computing environment. Through fast computation utility Cloud computing provides Computation as a Service (CaaS). Fast computation utility includes processing capability, size of memory, efficient algorithms etc.

Storage Utility: Instead of storing data at local storage device, service providers store them at storage device which is located at remote place. Storage utility consists of thousands of hard drives, flash drives, database servers etc. As computer systems are bound to fail over the period of time data redundancy is required. Due to Cloud's time variant service, storage utility needs to provide features like Cloud elasticity. Through storage, utility Cloud computing provides Storage as a Service (StaaS).

Communication Utility: It can also be called as Network Utility or Network as a Service (NaaS). Fast computation utility and storage utility cannot be thought without communication utility. Communication utility consists of physical (intermediate devices, hosts, sensors, physical communication link) and logical (bandwidth, delay, protocols, virtual communication link) resources. In Cloud computing, each and every service is provided through high speed Internet. So bandwidth and delay are most important from network point of view.

Power/Energy Utility: Nowadays researchers are doing a lot of research work on energy efficient techniques in Cloud computing. Energy cost can greatly be reduced by using power aware techniques. Due to thousands of data servers, power consumption is very high in Cloud computing. Cooling devices and UPS are at the center of these type of resources. They can also be considered as secondary resources.

Security Utility: Security is always a major issue in any computing area. Being Cloud user, we want highly reliable, trust-able, safe and secure Cloud services.

9.2   Issues in Resource Management

Issues in resource management are virtualization, network infrastructure management, data management, multi-tenancy, APIs, interoperability etc. These issues are briefly discussed as follows:

(i) Virtualization: There exists an issue of allocation of resources in virtualization networks (VNs). There exist several schemes for resource allocation in VNs. It is helpful when detail designs specifications and performance evaluation schemes for VNs are needed.

(ii) Network infrastructure management: managing multiple network components (bridges, hubs, switches etc.) leads to higher administrative costs. Hence, automated methods for managing it is required. It needs to scale effectively. Existing works discuss putting switches, routers, network interfaces and links into sleep modes in idle times; so that the excessive energy used by Internet can be saved.

(iii) Security and privacy are essential in Cloud environment dealing with sensitive data and code. Several security issues, such as: (a) authentication, (b) integrity, (c) data confidentiality and (d) non-repudiation should be considered to ensure adequate security.

(iv) To offer elastic pricing, charging and billing, metering of resources and its consumption is required. The issue here is to see that the users are charged only for the services they use for a specific period of time.

(v) Tools are required to support development, adaptation and usage of Cloud service. There are various tools that are available to do a specific task. The challenge is to select the tools that are effective, and give accurate results.

9.3   Solutions to Resource Management Issues

Solutions to resource management issues exists with respect to resource provisioning, resource allocation, resource adaptation and resource mapping. The existing literatures and challenges associated are discussed here.

9.3.1   Resource Provisioning

Cloud computing is a model for enabling convenient, on-demand network access to a shared pool of configurable computing resources (e.g., networks, servers, applications and services) that can be rapidly provisioned and released. Resource provisioning means the selection, deployment and run-time management of software (e.g., database server management systems, load balancers) and hardware resources (e.g., CPU, storage, and network) for ensuring guaranteed performance for applications.

Resource provisioning is an important and challenging problem in the large-scale distributed systems such as Cloud computing environments. There are many resource provisioning techniques, both static and dynamic each one having its own advantages and also some challenges. These resource provisioning techniques used must meet Quality of Service (QoS) parameters like availability, throughput, response time, security, reliability etc., and thereby avoiding Service-Level-Agreement (SLA) violation.

The real time situation of the Cloud computing environment is dynamic and complicated. There are many unexpected large changes occurring in the Cloud environment. The provisioning is based only on the user application objectives needs to be improvised for meeting all the above issues of the Cloud computing environment. Ant colony optimization (ACO) is used to optimize the result of provisioning by considering the characteristics of the Cloud environment. Many Artificial intelligence techniques have been used in research

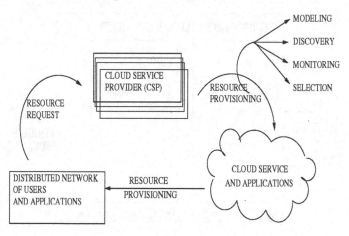

FIGURE 9.1
Resource Provisioning System in Cloud

for optimizations. Swarm intelligence is an approach to problem solving that takes inspiration from the social behaviors of insects and of other animals.

To implement a Resource Provisioning System (RPS) as shown in Figure 9.1, a Cloud Service Provider (CSP) deals with four concerns: Resource Modeling, Discovery, Monitoring and Selection. To model the resources in the environment is the first and foremost pillar of the provisioning model of a CSP. The discovery mechanism accounts for the resources ready to be leased for the users. For dynamic provisioning, an RPS needs to be aware of the real time status of the Cloud Resources. The selection of resources to be leased to a particular user is based on the status as monitored by the monitoring system.

The model discussed here tackles the problem of Multi-Objective Decision making for resource provisioning in two modules as shown in Figure 9.2. Resource Interest Score Evaluation (RISE) helps in relating the alternatives to the goal, and for evaluating Consistent Cloud Resource Interest scores for all alternative Cloud resources. The Optimal Resource Provisioning (ORP) provides optimal solution out of all the consistent options. The model has an ability to rank choices in order of their effectiveness in meeting conflicting objectives with efficiently detecting the inconsistent judgments. But as the decision problem is decomposed into a number of subsystems, pair wise comparisons are to be made under each subsystem.

The ORP optimizes the Cloud Resource Interest Score (CRIS) provided by the RISE module. The framework accounts various objectives and assigns consistent weightage to each influencing criteria. These influencing criterias vary according to the user requirements. Assigning weightage to the factors

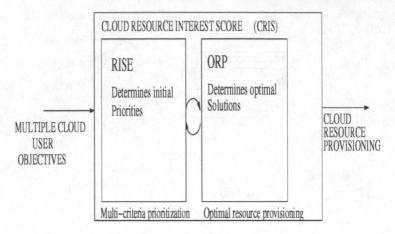

FIGURE 9.2
Optimized Framework for Resource Provisioning in Cloud

ensure that it meets multiple user application objectives. The criterias are divided into broad categories Task Expenses and Task Services. All the user requests are termed as Task which are to be handled by RPS. The expenses of the cost, a complete time and agility are structured under Task Expenses and Security, Assurance and Performance are under Task Services. All the alternatives are ranked according to these criterias. The lowest level of the hierarchy depicts the resource alternatives which are to be mapped to the appropriate alternative. The RISE provides CRIS by prioritizing the alternative according to these criterias. The chosen CRIS will be arranged using the ORP module to find the optimal provisioning.

The development of efficient service provisioning policies is among the major issues in Cloud research. The issues are to provide better quality of service in IaaS by provisioning the resources to the users or applications via load balancing mechanism, high availability mechanism, etc. In this context, game theoretic methods allows us to gain an in depth analytical understanding of the service provisioning problem. Earlier, game theory has been successfully applied to diverse problems such as Internet pricing, congestion control, routing and networking. Resource provisioning can encompass three dimensions: hardware resources, the software available on those resources and the time during which those resources must be guaranteed to be available. A complete resource provisioning model must allow resource consumers to specify requirements across these three dimensions, and the resource provider to efficiently satisfy those requirements.

A summary of some of the resource provisioning schemes is given in Table 9.1. Table 9.2 lists out the performance metrics of the resource provisioning schemes.

TABLE 9.1

Resource provisioning schemes

Name of the scheme	Functioning
Nash equilibrium approach using Game theory [2]	Run time management and allocation of IaaS resources considering several criteria such as the heterogeneous distribution of resources, rational exchange behaviors of Cloud users, incomplete common information and dynamic successive allocation.
OpenNebula(infrastructure manager) and Haizea (resource lease manager)[3]	Allow resource consumers to specify requirements across these three dimensions-hardware resources, the software available on those resources, and the time during which those resources must be guaranteed to be available for the resource provider to efficiently satisfy those requirements.
Resource pricing[4]	The provisioning procedure consists of two algorithms, one executed by the network and the other by individual users. The network offers resources freely to meet their desired quality based on their own traffic parameters and delay requirements. The network periodically adjusts resource prices based on user requests.
Network queuing model [5]	Presents a model based on a network of queues, where the queues represent different tiers of the application. The model sufficiently captures the behavior of tiers with significantly different performance characteristics such as session-based workloads, concurrency limits, and caching at intermediate tiers.
Prototype provisioning [6]	Employs the k-means clustering algorithm to automatically determine the workload mix and a queuing model to predict the server capacity for a given workload mix. A prototype provisioning system evaluate its efficiency on a laboratory Linux data center running the TPC-W web benchmark.

TABLE 9.2

Performance metrics for resource provisioning schemes

Schemes	Metrics				
	Reliability	Ease of Deployment	QoS	Delay	Control overhead
Nash equilibrium using Game theory [2]	High	Medium	High	Medium	High
OpenNebula(infrastructure manager) and Haizea (resource lease manager)[3]	High	High	High	High	High
Resource pricing [4]	Med	Medium	High	Medium	Medium
Network queuing model [5]	Medium	Low	Medium	Medium	Medium
Prototype provisioning[6]	Medium	Medium	Medium	High	High

9.3.2 Resource Allocation

In Cloud computing, Resource Allocation (RA) is the process of assigning available resources to the needed Cloud applications over the Internet. Resource allocation starves services if the allocation is not managed precisely.

Resource provisioning solves that problem by allowing the service providers to manage the resources for each individual module. Resource Allocation Strategy (RAS) is all about integrating Cloud provider activities for utilizing and allocating scarce resources within the limit of Cloud environment so as to meet the needs of the Cloud application. It requires the type and amount of resources needed by each application in order to complete a user job. The order and time of allocation of resources are also an input for an optimal RAS. An optimal RAS should avoid the following criteria as follows:

- Resource contention situation arises when two applications try to access the same resource at the same time.

- Scarcity of resources arises when there are limited resources.

- Resource fragmentation situation arises when the resources are isolated. There will be enough resources but not able to allocate to the needed application.

- Over-provisioning of resources arises when the application gets surplus resources than the demanded one

- Under-provisioning of resources occurs when the application is assigned with fewer numbers of resources than the demand.

Resource allocation has a significant impact in Cloud computing, especially in pay-per-use deployments where the number of resources are charged to application providers. The issue here is to allocate proper resources to perform the computation with minimal time and infrastructure cost. Proper resources are to be selected for specific applications in IaaS. Once the required types of resources are determined, instances of these resources are allocated to execute the task. Resource determination and allocation for each atomic task is managed by task modules.

Table 9.3 summarizes some of the resource allocation schemes. Table 9.4 lists out the performance metrics of the resource allocation schemes.

9.3.3   Resource Mapping

Mapping of virtual resources to physical resources has an impact on Cloud clients. Resource mapping is a system-building process that enables a community to identify existing resources and match those resources to a specific purpose. The issue here is to maximize Cloud utilization in IaaS by calculating the capacity of application requirements so that minimal Cloud computing infrastructure devices shall be procured and maintained. This can be achieved by using cognitive architecture that automatically builds a model of the machine's behavior based on prior training data.

In a Cloud computing environment, a logical network (i.e., a set of virtual machines) must be deployed on to physical network (servers). This requires mapping of VMs to physical resources. The mapping problem translates

TABLE 9.3

Resource allocation schemes

Name of the scheme	Functioning
Novel, non-intrusive method[23]	Proposes a novel, non-intrusive method for application and remoting protocol agnostic desktop responsiveness monitoring. Moreover, desktop workload usage enables to discover and leverage workload patterns to increased efficiency both in terms of desktop responsiveness and resource usage, is also highlighted.
Market-oriented how to resource allocation[24]	Considers the case of a single Cloud provider and address the question best match customer demand in terms of both supply and price in order to maximize the providers revenue and customer satisfactions while minimizing energy cost.
Intelligent multi-agent model[25]	Proposes an intelligent multi-agent model for resource virtualization (IMAV) to automatically allocate service resources suitable for mobile devices. It infers user's demand by analyzing and learning user context information. In addition, it allocates service resources according to use types so that users are able to utilize reliable service resources.
Mixed integer optimization techniques[26]	Applies a generic model for router power consumption model in a set of target network configurations and uses mixed integer optimization techniques to investigate power consumption, performance and robustness in static network design and in dynamic routing.
Energy-Aware Resource allocation[28]	Allocation is carried out by mimicking the behavior of ants, where ants are likely to choose the path identified as a shortest path, which is indicated by a relatively higher density of pheromone left on the path compared to other possible paths.

TABLE 9.4

Performance metrics for resource allocation schemes

Schemes	Metrics				
	Reliability	Ease of Deployment	QoS	Delay	Control overhead
Novel, non-intrusive method[23]	Medium	Medium	Medium	Medium	High
Market-oriented resource allocation[24]	Medium	High	Medium	High	High
Intelligent multi-agent model[25]	Medium	High	Medium	Low	Medium
Mixed integer optimization techniques[26]	Medium	Medium	High	High	Medium
Energy aware resource allocation[28]	High	Medium	High	Medium	Medium

virtual machines assignment onto physical servers and assigns flows in the physical network with bandwidth allocation so that requirements of logical communication can be met.

Modified Round Robin Algorithm for Resource Allocation

A modified round robin scheduling algorithm (Algorithm 9.1) is proposed for resource allocation in Cloud computing. It begins with the time equals to the time of first request, which changes after the end of first request. When a new request is added into the ready queue in order to be granted, the algorithm calculates the average of sum of the times of requests found in the ready queue including the new arrival request. This needs two registers: (i) SR: To store the sum of the remaining burst time in the ready queue, and (ii) AR: To store the average of the burst times by dividing the value found in the SR by the count of requests found in the ready queue. After execution, if request finishes its burst time, then it will be removed from ready queue or else it will move to the end of the ready queue. SR will be updated by subtracting the time consumed by this request. AR will be updated according to the new data.

Table 9.5 summarizes some of the resource mapping schemes. Table 9.6 lists out performance metrics of the resource mapping schemes.

TABLE 9.5
Resource mapping schemes

Name of the Scheme	Functioning
Mapping logical plane to underlying physical plane[49]	Presented a novel set of feasibility checks for node assignments based on graph cuts.
Symmetric mapping pattern[50]	Presents the symmetric mapping pattern, an architectural pattern for the design of resource supply systems through three functions: (1) Users and providers match and engage in resource supply agreements, (2) users place tasks on subscribed resource containers and (3) providers place supplied resource containers on physical resources.
Load-aware mapping[51]	Simplifies VM image management and reduce image preparation overhead by the multicast file transferring and image caching/reusing. Additionally, the Load-Aware Mapping, a novel resource mapping strategy, is proposed in order to further reduce deploying overhead and make efficient use of resources.
Minimum congestion mapping[52]	A framework for solving a natural graph mapping problem arising in Cloud computing. And applying this framework to obtain offline and online approximation algorithms for workloads given by depth-d trees and complete graphs.
Iterated local search based request partitioning[53]	A novel request partitioning approach based on iterated local search is introduced that facilitates the cost-efficient and on-line splitting of user requests among eligible Cloud Service Providers (CSPs) within a networked Cloud environment.

Algorithm 4 Algorithm 9.1: Modified Round Robin Algorithm

Inputs: Registers: SR, AR; Burst Time of Process (BT(P)), Time Queue (TQ);
Output: Executed Process P;
Description: Step 1: New request P arrives
Step 2: P Enters ready queue
Step 3: Update SR and AR
Step 4: Request P is loaded from ready queue into CPU queue to be executed
Step 5: While (Ready Queue! = NULL) do
Step 6: Ready Queue
Step 7: Process P
Step 8: Update SR & AR
Step 9: Load P
Step 10: end while
Step 11: If (Ready Queue = NULL) then
Step 12: TQ = BT (P)
Step 13: Update SR & AR
Step 14: else
Step 15: TQ = AVG (BT of all request in Ready Queue)
Step 16: Update SR & AR
Step 17: If (P terminated) then
Step 18: Update SR & AR
Step 19: else
Step 20: TQ = AVG (BT of all request in Ready Queue)
Step 21: Update SR & AR
Step 22: If (P terminated) then
Step 23: Update SR & AR
Step 24: else
Step 25: Return P
Step 26: Update SR & AR
Step 27: end if

TABLE 9.6

Performance metrics for resource mapping schemes

Schemes	Metrics				
	Reliability	Ease of Deployment	QoS	Delay	Control overhead
Mapping logical plane to underlying physical plane[49]	Medium	Med	High	Medium	High
Symmetric mapping pattern[50]	Medium	Medium	Medium	Medium	Medium
Load-aware mapping[51]	Medium	Medium	Medium	Medium	Medium
Minimum congestion mapping[52]	Medium	High	Medium	Low	Medium
Iterated local search based request partitioning[53]	High	Medium	Medium	High	High

TABLE 9.7

Resource adaptation schemes

Name of the Scheme	Functioning
Reinforcement learning guided control policy[54]	A framework that is a multi-input multi-output feedback control model-based dynamic resource provisioning algorithm which adopts reinforcement learning to adjust adaptive parameters.
Web-service based prototype[55]	Developed a fully functional web-service based prototype framework, and used it for performance evaluation of resources.
Mistral framework[56]	A framework that optimizes power consumption, performance benefits, and transient costs incurred by various adaptations.
OnTimeMeasure service[57]	Presents an application - adaptation case study that uses OnTimeMeasure-enabled performance in the context of dynamic resource allocation within thin-client based VMs
Virtual networks[58]	Proposes VN architecture in Cloud computing that can aggregate traffic isolation, improving security, and facilitating pricing.

9.3.4   Resource Adaptation

The primary reason for adapting Cloud computing from a user perspective is to move from the model of capital expenditure (CAPEX) to operational expenditure (OPEX) . Instead of buying IT resources like machines, storage devices etc. and employing personnel for operating, maintaining etc., a company pays another company (the "provider") for the actual resources used (pay-as-you-go). An important aspect of this is that a company no longer needs to overprovision its IT resources. It is typical today, when a company invests in its own resources, that the amount of resources invested in corresponds to the maximum amount of resources needed at peak times – with the result that much of these resources are not needed at all during regular periods.

Table 9.7 summarizes some of the resource adaptation schemes. Table 9.8 lists out the performance metrics of the resource adaptation schemes.

Summary

In this chapter, we surveyed research on resource management for Cloud environments. While working on this survey, we were surprised by the amount of recent results that we found, and the paper grew therefore larger than anticipated. To clarify the discussion and better place individual contributions in context, we outlined a framework for Cloud resource management, which lays the basis for the core of the chapter, the state-of-the-art survey. We concluded

TABLE 9.8

Performance metrics for resource adaptation schemes

Schemes	Metrics				
	Reliability	*Ease of Deployment*	*QoS*	*Delay*	*Control overhead*
Reinforcement learning guided control policy[54]	High	Medium	High	Medium	High
Web-service based prototype[55]	Medium	High	Medium	Medium	High
Mistral framework[56]	Medium	Low	Medium	High	High
OnTimeMeasure service[57]	Medium	Medium	High	Medium	Low
Virtual networks[58]	Medium	Medium	Medium	High	High

the chapter with a set of fundamental research challenges, which we hope will encourage new activities in this fascinating and timely field.

Keywords

Resource mangement	Resource provisioning
Resource allocation	Resource adaptation

Objective type questions

1. A solution can be partitioned into groups of logic that can be designated for both frontend and backend instance invocation so as to optimize runtime execution and billing.

(a) Pay-per-use monitor

(b) SLA Management System

(c) Ready-made environment

(d) Resource Replication

2. The multi-device broker mechanism is used to facilitate runtime data transformation so as to make a Cloud service accessible by a wider range of Cloud service consumer programs and devices.

(a) Virtual Server

(b) Multi-Devices Broker

(c) Failover System

(d) Hypervisor

3. The audit monitor mechanism is used to collect audit tracking data for networks and IT resources in support of, or dictated by, regulatory and contractual obligations.

(a) Pay-per-use monitor

(b) SLA Monitor

(c) Audit Monitor

(d) Cloud Usage Monitor

4. Different Cloud provider vendors have different names for service agents that act as automated scaling listeners.

(a) Load Balancer   (b) Audit Monitor

(c) Resource Cluster   (d) Automated Scaling Listener

5. The data collected by the pay-per-use monitor is processed by a billing management system that calculates the payment fees.

(a) SLA Monitor   (b) Pay-per-use monitor

(c) Cloud Usage Monitor   (d) Audit Monitor

6. This is essentially a shopping portal that allows Cloud consumers to search an up-to-date list of Cloud services and IT resources that are available from a Cloud provider (usually for lease). The Cloud consumer submits its chosen items to the Cloud provider for provisioning.

(a) Failover System   (b) SLA Management System

(c) Remote Administration System   (d) Resource Management System

7. Monitoring operational conditions of IT resources is ?

(a) Resource Cluster   (b) Billing Management System

(c) SLA Management System   (d) Resource Management System

8. As an alternative to caching state data in memory, software programs can offload state data to the database in order to reduce the amount of run-time memory they consume. State management databases are commonly used by Cloud services, especially those involved in long-running runtime activities.

(a) Resource Management System   (b) SLA Management System

(c) Billing Management System   (d) State Management Database

9. A form of virtualization software that emulates a physical server and is used by Cloud providers to share the same physical server with multiple Cloud consumers by providing Cloud consumers with individual virtual server instances.

(a) SLA Monitor   (b) Failover System

(c) Virtual Server   (d) Load Balancer

10. These systems are commonly used for mission-critical programs or for reusable services that can introduce a single point of failure for multiple applications.

(a) SLA Management System   (b) Failover System

(c) Virtual Server   (d) Hypervisor

11. Which of the following is a core management feature offered by most Cloud management service products ?

(a) Support of different Cloud types

(b) Creation and provisioning of different types of Cloud resources, such as machine instances, storage, or staged applications

(c) Performance reporting including availability and uptime, response time, resource quota usage, and other characteristics

(d) All of the above

12. Point out the correct statement:

(a) Eucalyptus and Rackspace both use Amazon EC2 and S3 services

(b) The RightScale user interface provides real-time measurements of individual server instances

(c) RightScale server templates and the Rightscript technology are highly configurable and can be run under batch control

(d) All of the above

13. Which of the following is open and both hypervisor and processor-architecture-agnostic ?

(a) Dell Scalent

(b) CloudKick

(c) Elastra

(d) All of the above

14. Point out the wrong statement :

(a) All applications benefit from deployment in the Cloud

(b) With Cloud computing, you can start very small and become big very fast

(c) Cloud computing is revolutionary, even if the technology it is built on is evolutionary

(d) All of the above

15. Which of the following monitors the performance of the major Cloud-based services in real time in Cloud Commons ?

(a) CloudWatch

(b) CloudSensor

(c) CloudMetrics

(d) All of the above

Objective type questions -answer

1:c 2:b 3:c 4:d 5:b 6:c 7:d 8:d 9:c 10:b 11:c 12:d 13:d 14:a
15:b

Review questions

1. What is an AMI? How do we implement it?
2. Explain vertical scaling in an Amazon instance.
3. Mention the key components of AWS.
4. What are the characteristics of Cloud architecture that differs from traditional Cloud architecture.
5. Mention the different datacenters deployment of Cloud computing.
6. List the platforms which are used for large scale Cloud computing.
7. List the challenges w.r.t resource provisioning in Cloud computing.
8. List the challenges w.r.t resource allocation in Cloud computing.
9. List the challenges w.r.t resource adaptation in Cloud computing.
10. List the challenges w.r.t resource mapping in Cloud computing.

Critical thinking questions

1. Analyze the benefits and the problems posed by the four approaches for the implementation of resource management policies: control theory, machine learning, utility-based, market-oriented.

2. Can optimal strategies for the five classes of policies, admission control, capacity allocation, load balancing, energy optimization and QoS guarantees be actually implemented in a Cloud? The term "optimal" is used in the sense of control theory. Support your answer with solid arguments. Optimal strategies for one class may be in conflict with optimal strategies for one or more of the other classes. Identify and analyze such cases.

3. Analyze the relationship between the scale of a system and the policies and the mechanisms for resource management. Consider also the geographic scale of the system in your arguments.

4. Multiple controllers are probably necessary due to the scale of the Cloud. Is it beneficial to have system and application controllers? Should the controllers be specialized for example, some to monitor performance, others to monitor power consumption? Should all the functions we want to base the resource management policies on be integrated in a single controller and one controller be assigned to a given number of servers, or to a geographic region? Justify your answers.

5. Analyze the challenges of transition to $IPv6$. What will be in your view the effect of this transition on Cloud computing?

6. Use the start-time fair queuing (SFQ) scheduling algorithm to compute the virtual startup and the virtual finish time for two threads a and b with weights $w\_a = 1$ and $w\_b = 5$ when the time quantum is $q = 15$ and thread b blocks at time $t = 24$ and wakes up at time $t = 60$. Plot the virtual time of the scheduler function of the real time.

Bibliography

[1] Sunilkumar. S. Manvi, Gopal Krishna Shyam, "Resource management for Infrastructure as a Service (IaaS) in Cloud computing: A Survey", *Journal of Network and Computer Applications, Elsevier*, Vol. 41, No. 1, pp. 424-440, 2014.

[2] F. Teng, F. Magoules, "A new game theoretical resource allocation algorithm for Cloud computing", *Proceedings of 1st International Conference on Advances in Grid and Pervasive Computing*, Vol.6, no.4, 2019, pp. 321-330.

[3] Borja Sotomayor, Rubn S. Montero, Ignacio M. Llorente, Ian Foster, "An Open source solution for virtual infrastructure management in private and hybrid Clouds", *Proceedings of the IEEE International Conference on Internet Computing*, Vol. 10, no.6, 2019, pp. 78-89.

[4] Murray Hill, Varaiya.P, "An algorithm for optimal service provisioning using resource pricing", *Proceedings of 13th IEEE International Conference on Networking for Global Communications*, Vol.1, no. 2, 2009 , pp. 368-373 .

[5] B. Urgaonkar, G. Pacifici, P. Shenoy, M. Spreitzer, A. Tantawi, "An analytical model for multi-tier Internet services and its applications", *Proceedings of the ACM SIGMETRICS International Conference on Measurement and Modeling of Computer Systems*, Vol. 33, no.5, 2010, pp. 291-302.

[6] R. Singh, U. Sharma, E. Cecchet, P. Shenoy, "Autonomic mix-aware provisioning for non-stationary data center workloads", *Proceedings of 7th IEEE International Conference on Autonomic Computing and Communication*, Vol.8, no.4, 2010, pp 24-31.

[7] S. Cunningham, G. Holmes, "Developing innovative applications of machine learning", *Proceedings of Southeast Asia Regional Computer Confederation Conference*, Vol. 6, no. 4, 2011, pp. 67-76.

[8] Vijayakumar.S, Qian Zhu, Agrawal. G, "Automated and dynamic application accuracy management and resource provisioning in a Cloud environment", *Proceedings of 11th IEEE/ACM International Conference on Grid Computing*, Vol. 5, no.6 , 2010, pp. 33-40.

[9] Daniel Warneke, Odej Kao, "Exploiting dynamic resource allocation for efficient parallel data processing in the Cloud", *IEEE Transactions on Parallel and Distributed Systems*, Vol. 6, no.4, 2011, pp. 34-48.

[10] G. Juve and E. Deelman, "Resource provisioning options for large-scale scientific workflows", *Proceedings of 4th IEEE International Conference on e-Science*, Vol. 13, no.7, 2012, pp. 608-613.

[11] Z. Huang, C. He, J. Wu, "Architecture design, and implementation", *Proceedings of 11th International Conference on Parallel and Distributed Systems*, Vol.10, no.7, 2011, pp. 65-75.

[12] Y. Jie, Q. Jie, L. Ying, "A profile-based approach to just-in- time scalability for Cloud applications", *Proceedings of IEEE International Conference on Cloud Computing*, Vol.3, no.2, 2018, pp. 101-118.

[13] Y. Kee and C. Kesselman, "Grid resource abstraction, virtualization, and provisioning for time-target applications", *Proceedings of IEEE International Symposium on Cluster Computing and the Grid*, Vol.11, no.3, 2011, pp. 199-203.

[14] A. Filali, A.S. Hafid, M. Gendreau, "Adaptive resources provisioning for Grid applications and services", *Proceedings of IEEE International Conference on Network Communications*, Vol. 2, no.14, 2009, pp. 607-614.

[15] D. Kusic and N. Kandasamy, "Risk-Aware limited lookahead control for dynamic resource provisioning in enterprise computing systems", *Proceedings of IEEE International Conference on Autonomic Computing*, Vol.10, no.3, 2010, pp. 337-350.

[16] K. Miyashita, K. Masuda, and F. Higashitani, "Coordinating service allocation through flexible reservation", *IEEE Transactions on Services Computing*, Vol.5, no.2, 2011, pp. 65-69.

[17] J. Chen, G. Soundararajan, C. Amza, "Autonomic provisioning of back-end databases in dynamic content web servers", *Proceedings of IEEE International Conference on Autonomic Computing*, Vol.7, no.3, 2011, pp. 231-242.

[18] Ruben S. Montero, Rafael Moreno-Vozmediano, Ignacio M. Llorente, "An elasticity model for high throughput computing clusters", *Journal of Parallel and Distributed Computing archive*, Vol. 71, no. 6, 2018, pp. 750-757.

[19] Xiangzhen Kong, Chuang Lin, Yixin Jiang, Wei Yan, Xiaowen Chu, "Efficient dynamic task scheduling in virtualized data centers with fuzzy prediction", *Journal of Network and Computer Applications*, Vol. 34, no. 4, 2011, pp. 1068-1077.

[20] Vianna E, "Modeling performance of the hadoop online prototype", *International Symposium on Computer Architecture*, Vol. 53, no. 1, 2012, pp. 72-77.

[21] Xie J, "Improving Map Reduce performance through data placement in heterogenous Hadoop clusters", *IEEE International Symposium on Parallel and Distributed Processing*, Vol. 66, no. 10, 2016, pp. 1322-1337.

[22] Amazon EC2, *http://aws.amazon.com/ec2/*, Accessed on 09.07.2019.

[23] R. Bhowmik, A. Kochut, K. Beaty, "Managing responsiveness of virtual desktops using passive monitoring", *Proceedings of IEEE Integrated Network Management Symposium*, Vol.28, no.4, 2010, pp. 45-51.

[24] Qi Zhang, Quanyan Zhu, Raouf Boutaba, "Dynamic resource allocation for spot markets in Cloud computing environment", *Proceedings of Fourth IEEE International Conference on Utility and Cloud Computing*, Vol. 10, no. 6, 2011, pp. 177-185.

[25] Myougnjin Kim , Hanku Lee , Hyogun Yoon, Jee-In Kim, HyungSeok Kim, "IMAV: An Intelligent Multi-Agent model based on Cloud computing for resource Virtualization", *Proceedings of International Conference on Information and Electronics Engineering*, Vol 6, no.5, 2011, pp. 30-35.

[26] J. Chabarek, J. Sommers, P. Barford, C. Estan, D.T.S. Wright, "Power awareness in network design and routing", *Proceedings of the 27th IEEE Conference on Computer Communications*, Vol.20, no.4, 2010, pp.457-465.

[27] L. Chiaraviglio, I. Matta, "GreenCoop: Co-operative green routing with energy efficient servers", *Proceedings of the 1st ACM International Conference on Energy-Efficient Computing and Networking*, Vol.3, no.2, 2011, pp. 191-194.

[28] L. Chiaraviglio, I. Matta, "Resource allocation using energy-efficient servers", *Proceedings of the 1st ACM International Conference on Energy-Efficient Computing and Networking*, Vol.6, no.4, 2010, pp. 54-67.

[29] Carlo Batini, Simone Grega, Andrea Maurino, "Optimal enterprise data architecture", *Proceedings of the 19th ACM International Symposium on High Performance Distributed Computing*, Vol.8, no.4, 2011, pp. 541-547.

[30] Dan Upton, "Enabling efficient online profiling of homogeneous and heterogeneous multicore systems", *Proceedings of Symposium on Microarchitecture*, Vol.3, no.4, 2010, pp. 22-31.

[31] K.Hatakeyama, Y.Osana, M.Tanabe, S.Kuribayashi, "Proposed congestion control method reducing the size of required resource for all-IP",
*Proceedings of IEEE Pacific Rim Conference on Communications, Com-
puters and Signal Processing*, Vol.11, no.3, 2009, pp. 124-135.

[32] T.Yoshino, Y.Osana and S.Kuribayashi, "Evaluation of congestion con-
trol methods for joint multiple resource allocation", *Proceeding of 13th
International Conference on Network-based Information Systems*,
Vol.10, no.3, 2010, pp. 16-18.

[33] Ming Mao, Marty Humphrey, "Auto-scaling to minimize cost and meet
application deadlines in Cloud workflows", *Proceedings of International
Conference for High Performance Computing, Networking, Storage and
Analysis*, Vol. 37, no.4, 2012, pp. 337-348.

[34] A. Ruiz-Alvarez, M. Humphrey, "A model and decision procedure for
data storage in Cloud computing", *Proceedings of the IEEE/ACM International Symposium on Cluster, Cloud, and Grid Computing*, Vol.
12, no.4, 2012, pp. 50-52.

[35] Chandrashekhar S. Pawar, R.B.Wagh, "A review of resource allocation
policies in Cloud computing", *Proceedings of National Conference on
Emerging Trends in Information Technology*, World Journal of Science
and Technology, Vol. 2, no. 3, 2012, pp. 165-167.

[36] A. Ruiz-Alvarez, M. Humphrey, "An automated approach to Cloud storage service selection", *Proceedings of the 2nd Workshop on Scientific
Cloud Computing*, Vol. 10, no.2, 2011, pp. 44-56.

[37] Yazir Y.O, Matthews C, Farahbod R, Neville S, Guitouni A, Ganti S,
Coady Y, "Dynamic resource allocation based on distributed multiple
criteria decisions in computing Cloud", *Proceedings of 3rd International
Conference on Cloud Computing*, Vol. 28, no.1, 2010, pp. 91-98.

[38] Arkaitz Ruiz Alvarez, Marty Humphrey, "An automated approach to
Cloud storage service selection", *Proceedings of the 2nd international
workshop on Scientific Cloud computing*, Vol. 8, no. 3, 2011, pp. 39-48.

[39] Z. Hill, M. Humphrey, "CSAL: A Cloud Storage Abstraction Layer to
enable portable Cloud applications", *Proceedings of 2nd IEEE International Conference on Cloud Computing Technology and Science*, Vol. 12,
no.5, 2011, pp. 67-72.

[40] Kevin Lai, Lars Rasmusson, Eytan Adar, Li Zhang, Bernardo A, "Ty-
coon: An implementation of a distributed, market-based resource allocation system", *Journal of Multiagent and Grid Systems*, Vol.1, no.3,
2005, pp. 169-182.

[41] Rajkumar Buyya, Chee Shin Yeo, and Srikumar Venugopal, "Market-Oriented Cloud computing: Vision, hype, and reality for delivering IT services as computing utilities", *Journal of Future Generation Computer Systems archive*, Vol. 25, no. 6, 2009, pp. 599-616.

[42] Shin-ichi Kuribayashi, "Optimal joint multiple resource allocation method for Cloud computing environments", *Journal on Research and Reviews in Computer Science*, Vol.2, no.1, 2011, pp. 155-168.

[43] Srikumar Venugopal, Xingchen Chu, and Rajkumar Buyya, "A negotiation mechanism for advance resource reservation using the Alternate Offers Protocol", *Journal on Future Generation Computer Systems*, Vol.25, no. 6, 2009, pp. 599-616.

[44] Neha Tyagi, Rajesh Pathak, "Negotiation for resource allocation for multiple processes Infrastructure as a service Cloud", *Journal on Advances in Engineering Research*, Vol.10, no.2, 2011, pp. 65-71.

[45] Elena Apostol, Valentin Cristea, "Policy based resource allocation in Cloud systems", *Proceedings of 2011 International Conference on P2P, Parallel, Grid, Cloud and Internet Computing*, Vol.7, no.1, 2011, pp. 165-172.

[46] Gokul Soundararajan, Daniel Lupei, Saeed Ghanbar, "Dynamic resource allocation for database servers running on virtual storage", *Proceedings of 7th USENIX Conference on File and Storage Technologies*, Vol.6, no.2, 2011, pp. 71-84.

[47] Norman Bobroff, Andrzej Kochut, Kirk Beaty, "Dynamic placement of virtual machines for managing SLA violations", *Proceedings of the 8th USENIX conference on File and storage technologies*, Vol.10, no.3, 2010, pp. 20-21.

[48] Jianzhe Tai, Juemin Zhang, Jun Li, Walced Meleis, Ningfang Mi, "Adaptive Resource Allocation for Cloud Computing Environments under Bursty Workloads", *Proceedings of International Conference on Communications*, Vol.2, no.2, 2011, pp. 55-68.

[49] Yun Hou, Murtaza Zafer, Kang-won Lee, Dinesh Verma, Kin K. Leung, "On the mapping between logical and physical topologies", *Proceedings of IEEE Conference on COMSNETS*, Vol.6, no.4, 2009, pp. 67-75.

[50] Xavier Grehant, Isabelle Demeure, "Symmetric mapping: An architectural pattern for resource supply in Grids and Clouds", *Proceedings of 2011 IEEE Conference on IPDPS*, Vol.11, no.4, 2011, pp. 1-8.

[51] Yang Chen, Tinnyu Wo, Jianxin Li, "An efficient resource management system for on-line virtual cluster provision ", *Proceedings of 2009 IEEE International Conference on Cloud Computing*, Vol.12, no.3, 2009, pp. 72-79.

[52] Nikhil Bansal, Kang-Won Lee, Viswanath Nagarajan, and Murtaza Zafer, "Minimum congestion mapping in a Cloud", *Proceedings of 2011 IEEE Conference on PODC*, Vol.6, no.5, 2011, pp. 267-276.

[53] A. Leivadeas, C. Papagianni, and S. Papavassiliou, "Efficient resource mapping framework over networked Clouds via Iterated Local Search based Request Partitioning", *IEEE Transactions on Parallel and Distributed Systems, Special section on Cloud computing*, Vol.11, no.3, 2011, pp. 521-525.

[54] Qian zhu,Gagan Agrawal, "Resource provisioning with budget constraints for adaptive applications in Cloud environments", *Proceedings of HPDC*, Vol.8, no.3, 2010, pp. 304-307.

[55] Ta Nguyen Binh Duong, Xiaorong Li, Rick Siow Mong Goh, "A framework for dynamic resource provisioning and adaptation in IaaS Clouds",*Proceedings of the 2011 IEEE Third International Conference on Cloud Computing Technology and Science*, Vol.5, no.4, 2009, pp. 312-319.

[56] Gueyoung Jung,Matti A. Hiltunen, Kaustubh R. Joshi, Richard D. Schlichting, Calton Pu, "Mistral: Dynamically managing power, performance, and adaptation cost in Cloud infrastructures", *Proceedings of the 2010 IEEE 30th International Conference on Distributed Computing Systems*, Vol.11, no.5, 2010, pp. 18-31.

[57] Prasad Calyam, Munkundan Sridharan, Yingxiao Xu, Kunpeng Zhu, Alex Berryman, Rohit Patal, "Enabling performance intelligence for application adaptation in the future Internet", *Journal on Communication and Networks*, Vol. 13, no.6, 2019, pp. 67-78.

[58] Carlos R. Senna, Milton A. Soares Jr, Luiz F. Bittencourt, Edmundo R. M. Madeira, "An architecture for adaptation of virtual networks on Clouds", *Proceedings of Network Operations and Management Symposium*, Vol.13, no.7, 2011, pp. 1-8.

[59] Schubert, "The future of computing. opportunities for European Cloud computing beyond 2020", *Available from http://bit.ly/b7faxz*, Accessed 26/07/2019.

[60] Sclater, "Cloud computing in education", *Policy Brief on Unesco Institute for Information Technology in Education*, Vol.11, no.5, 2011, pp.78-81.

[61] Charalambous.T, "Decision and Control", *Proceedings of 49th IEEE Conference on CDC*, Vol.28, no.11, 2010, pp. 3778-3783.

[62] Paul Marshall, Kate Keahey, Tim Freeman, "Elastic Site: Using Clouds to elastically extend site resources", *Proceedings of 10th IEEE/ACM International Conference on Cluster, Cloud and Grid Computing*, Vol.13, no.5, 2010, pp. 43-52.

[63] B. Raghavan, K. Vishwanath, S. Ramabhadran, K. Yocum, A. C. Snoeren, "Cloud control with distributed rate limiting", *Proceedings of the 2007 conference on Applications, Technologies, Architectures, and Protocols for Computer Communications*, Vol.17, no.4, 2009, pp. 337-348.

10

Cloud Computing Development Tools

Learning Objectives

After reading this chapter, you will be able to:

- Decide on mechanisms to procure infrastructure on the storage, compute, database and servers through the Cloud

- Describe features that all enterprises should look for when determining the best Cloud-based development tool

- Summarize automation of development tools, integration tools and user-interface tools via the Cloud

Tools that enable a developer to build and deploy an application in Cloud include KVM, DeltaCloud, Eucalyptus, OpenStack and Apache CloudStack Community Edition among others. They are discussed as follows.

10.1   Kernel Virtual Machine (KVM)

Virtualization technology for servers in the x86 family of CPUs has been attracting attention in recent years for various reasons. Server virtualization itself is a technology that has been around for some time, and the provision of Intel Virtualization Technology (Intel VT) and AMD-Virtualization (AMD-V) support functions in Intel and AMD CPUs, respectively, has provided developers with an environment that can achieve virtualization relatively inexpensive at practical levels of performance using x86 hardware. Various types of software for achieving server virtualization have also appeared.

Amidst these developments, the kernel based virtual machine (KVM) has rapidly come to the forefront as a server virtualization function provided as open source software (OSS). KVM, which was designed assuming use of the Intel VT-x note or AMD-V function, achieves virtualization using a relatively simple structure.

KVM is a full virtualization solution for Linux on x86 hardware containing virtualization extensions (Intel VT or AMD-V). It is an open source software.

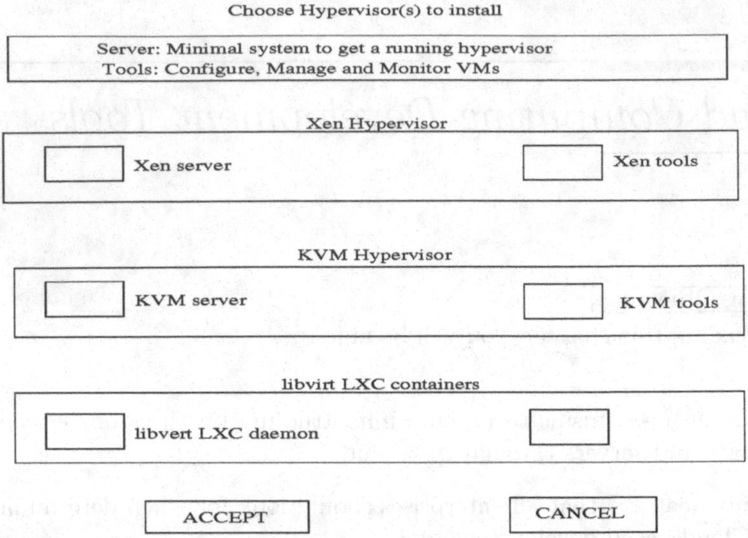

Choose Hypervisor(s) to install

Server: Minimal system to get a running hypervisor Tools: Configure, Manage and Monitor VMs

Xen Hypervisor

	Xen server		Xen tools

KVM Hypervisor

	KVM server		KVM tools

libvirt LXC containers

	libvert LXC daemon		

ACCEPT		CANCEL

FIGURE 10.1
Hypervisors

The kernel component of KVM is included in mainline Linux, as of 2.6.20. The userspace component of KVM is included in mainline QEMU, as of 1.3. The KVM components are discussed next.

KVM Components

Quick Emulator (QEMU) is a generic and open source machine emulator and virtualizer. QEMU itself is launched by entering a simple command of the same name via a character user interface (CUI) in Linux operating system. For example, if the process status (ps) command is entered while a guest system is running, the status of QEMU execution is displayed. The command options "-m 1024" and "-smp 1" indicate the memory capacity and the number of CPUs, respectively, of the guest system. It is not necessarily easy to specify all of these options directly with the QEMU command. For this reason, a graphical user interface (GUI), referred to as "virt-manager," has been prepared for Red Hat Enterprise Linux (RHEL) as well as for other operating systems to enable the user to operate and manage one or more guest systems.

Figure 10.1 shows KVM hypervisor installation in the OSs. In addition to such GUIs, virt-manager can also take the form of a CUI called "virsh," which can also be used to operate guest systems. The overall structure of KVM, from the GUI to the Linux kernel, includes five main components given as follows:

QEMU: An emulator that interacts with the KVM kernel module and executes many types of guest-system processing such as I/O; one QEMU process corresponds to one guest system.

libvirt: A tool-and-interface library common to server virtualization software supporting Xen, VMware ESX/GSX, and, of course, QEMU/KVM.

virt-manager: A GUI/CUI user interface used for managing VMs; it calls VM functions using libvirt.

KVM kernel module: In a narrow sense, KVM is a Linux kernel module; it handles VM Exits from guest systems and executes VM Entry instructions.

Linux kernel: Since QEMU runs as an ordinary process, scheduling of the corresponding guest system is handled by the Linux kernel itself.

10.2   DeltaCloud

With the ongoing ascent in Cloud figuring, most Cloud suppliers have offered their own APIs, which means Cloud clients pursue the administrations of different providers to the detriment of having the option to move effectively to different providers at a later stage. Apache DeltaCloud addresses this issue by offering a normalized API definition for framework as an assistance (IaaS) Clouds with drivers for a scope of various Clouds.

DeltaCloud is an application programming interface (API) developed by Red Hat and the Apache Software Foundation that abstracts differences between Cloud computing implementations. It was announced on September 3, 2009. The purpose of DeltaCloud is to provide one unified REST-based API that can be used to manage services on any Cloud. Each particular Cloud is controlled through an adapter called a "driver". As of June 2012, drivers exist for the following Cloud platforms: Amazon EC2, Fujitsu Global Cloud Platform, GoGrid, OpenNebula, Rackspace, RHEV-M, RimuHosting, Terremark and VMware vCloud. Next to the "classic" front-end, it also offers CIMI and EC2 front-ends.

DeltaCloud is used in applications such as Aeolus to prevent the need to implement Cloud-specific logic. Fig. 10.2 shows DeltaCloud betweem HTTP client and IaaS Cloud providers. Client-server is a relationship in which one program (the client) requests a service or resource from another program (the server). The DeltaCloud API is designed as a RESTful web service and comes

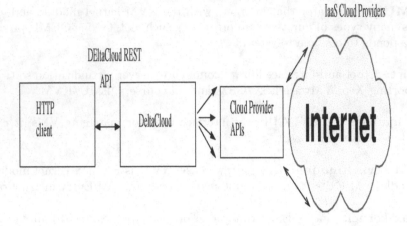

FIGURE 10.2
DeltaCloud

with client libraries for all major programming languages. Additional drivers for accessing further public or private Clouds can be created with minimal effort.

10.3   Eucalyptus

Eucalyptus represents Elastic Utility Computing Architecture for Linking Your Programs To Useful Systems. It is a powerful open source structure that gives a platform to private Cloud implementations on computer systems. It executes IaaS based arrangements in private and hybrid Clouds. Eucalyptus gives a platform to a single interface with the goal that clients can figure the assets accessible in private Clouds and the assets accessible remotely via public Cloud. It is structured dependent on extensible and and modular architecture for Web services. It likewise executes the business standard Amazon Web Services (AWS) API. This facilitates in exporting of large number of APIs for users by means of various tools.

10.3.1   Architecture

The architectural components of Eucalyptus include the following.

Images: Software modules, configuration, applications or software deployed in the Eucalyptus Cloud.

Instances: Images are called instances. The controller decides how to allocate memory and provide all resources to the instances.

Networking: This comprises three modes: the managed mode for the local network of instances including security groups and IP addresses, the system mode for MAC address allocation and attaching the instance network interface to the physical network via Node controller (NC) bridge, and the static mode for IP address allocation to instances.

Access control: Every use of Eucalyptus is assigned an identity and these identities are grouped together for access control.

Elastic block storage: This provides block-level storage volumes that are attached to an instance.

Auto scaling and load balancing: This creates and destroys instances automatically based on requirements.

Eucalyptus users have the capacity to monitor and execute all virtual machine deployed in their environment in an easy, flexible modular fashion as shown in Figure. 10.3. Cloud controller is one or more nodes, that control the Cloud operations. In contrast to Cloud worker nodes (e.g. compute nodes), Cloud controller maintains a high level view of the Cloud resources and provides unified point for Cloud management. All of the user's request (e.g. launch an instance) firstly goes into Cloud controller node and then according to the request, it sends to the other nodes (e.g. compute, storage or network nodes) to complete the operation.

A Node Controller (NC) executes on every node that is part of the Cloud infrastructure and designated to host a virtual machine instance. The NC answers the queries from the Cluster Controller using the system APIs of the operating system of the node and the hypervisor to get system information, such as number of cores, memory size, available disk space, state of the VM instance, etc.

10.3.2 Components

The following components make up the Eucalyptus Cloud.

Cloud controller: This is the main controller that manages the entire platform and provides a Web and EC2 compatible interface. It performs scheduling, resource allocation and accounting.

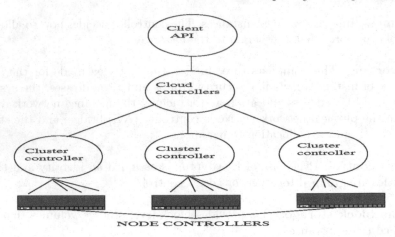

FIGURE 10.3
Eucalyptus

Walrus: Works like AWS S3 (Simple Storage Service). It offers persistent storage to all virtual machines in the Cloud and can be used as a simple HTTP put/get Solution as a Service.

Cluster controller: Manages VM execution and SLAs. Communicates with the storage controller and node controller, and is the front end for the cluster.

Storage controller: Communicates with the cluster controller and node controller, managing all block volumes and snapshots to the instances within the specific cluster.

VMware broker: Provides an AWS-compatible interface for VMware environments and physically runs on the cluster controller.

Node controller: Hosts all virtual machine instances and manages the virtual network endpoints.

10.4   OpenStack

OpenStack is a free and open-source platform for Cloud computing, deployed as IaaS, whereby virtual servers and different assets are made accessible to clients. The software platform comprises of related parts that control diverse, multi-vendor equipment pools of processing, storage, and networking resources through a data center. Clients either oversee it through an web-based dashboard, or through RESTful web services. OpenStack started in 2010 as a joint project of Rackspace Hosting and NASA.

Ecosystem: OpenStack comprises 6 core services and 13 optional services. Some of them are listed and discussed below.

Keystone: This is the authorisation and authentication component of OpenStack. It handles the creation of the environment and is the first component installed.

Nova: This manages the computing resources of the OpenStack Cloud. It takes care of instance creation and sizing as well as location management.

Neutron: This is responsible for creating virtual networks within the OpenStack Cloud. This module creates virtual networks, routers, sub-nets, firewalls, load balancers, etc.

Glance: This manages and maintains server images. It facilitates the uploading and storing of OpenStack images.

Cinder: This provides block storage, as a service, to the OpenStack Cloud. It virtualises pools of block storage devices and provides end users with a self-service API to request and consume resources without requiring any storage knowledge.

Swift: This comprises storage systems for objects and files. It makes scaling easy, as developers do not have to worry about the system's capacity. It allows the system to handle the data backup in case of any sort of machine or network failure.

Horizon: This is the dashboard of OpenStack. It provides a GUI interface to users to access all components of OpenStack via the API and enables administrators to use the OpenStack Cloud.

Ceilometer: This provides telemetry services, including billing services to individual users of the Cloud.
Heat: This is the orchestration component of OpenStack, allowing developers to define and store the requirements of a Cloud application in a file.

OpenStack can be deployed using the following models.

OpenStack-based public Cloud: In this case, the vendor provides a public Cloud computing system based on the OpenStack project.

On-premise distribution: Here, the clients download and install the distribution within their customised network.

Hosted OpenStack private Cloud: In this case, the vendor hosts the OpenStack based private Cloud, including the hardware and software.

OpenStack-as-a-Service: Here, the vendor hosts OpenStack management Software as a Service.

Appliance based OpenStack: This covers compatible appliances that can be added into the network provided by Nebula.

10.5   SaltStack

SaltStack or Salt is open source Cloud infrastructure configuration management software written in Python and supports the Infrastructure as a Code approach for Cloud management. It provides a powerful interface for Cloud hosts to interact and interface. Salt is designed to allow users to explicitly target and issue commands to multiple machines directly. It is based around the idea of a master, which controls one or more minions. Commands are normally issued from the master to a target group of minions. These minions execute the tasks specified in the commands and return the resulting data back to the master. Communication between a master and minions occurs over the ZeroMQ message bus.

SaltStack modules communicate with the supported minion operating systems. The Salt Master runs on Linux by default, but any operating system can be a minion, and currently Windows, VMware, vSphere and BSD UNIX variants are well supported. The Salt Master and the minions use keys to communicate. When a minion connects to a master for the first time, it automatically stores keys on the master. SaltStack also offers Salt SSH, which provides agentless systems management.

Architecture: The components that make up SaltStack architecture are: (i) SaltMaster: This send commands and configurations to Salt slaves, (ii) Salt-Minions: These receive commands and configurations from SaltMaster, (iii) Executions: These are modules and ad hoc commands, (iv) Formulas: These are used for installing a package, configuring and starting a service, and setting up users and permissions, (v) Grains: These offer an interface to minions, (vi) Pillar: This generates and stores sensitive information with regard to minions, (vii) Top file: This matches Salt states and pillar data to Salt minions, (viii)

Runners: These perform tasks like job status and connection status, and read data from external APIs, (ix) Returners: These are responsible for triggering reactions when events occur, and (x) SaltCloud: This is an interface to interact with Cloud hosts.

10.6   Apache CloudStack

Apache CloudStack is open source software especially designed to deploy and manage large networks of virtual machines to create highly scalable, available and secured IaaS based computing platforms. It is multi-hypervisor and multi-tenant, and provides an open, flexible Cloud orchestration platform delivering reliable and scalable private and public Clouds. CloudStack provides a Cloud orchestration layer that automates the creation, provisioning and configuration of IaaS components. It supports all the popular hypervisors: VMware, KVM, Citrix XenServer, Xen Cloud Platform, Oracle VM server and Microsoft Hyper-V.

Architecture: Apache CloudStack deployment involves two parts: the management server and the resources to be managed, as shown in Fig. 10.4. Two machines are required to deploy Apache CloudStack, the CloudStack management server and the Cloud infrastructure server.

1. The management server: This orchestrates and allocates the resources during Cloud deployment. It runs on a dedicated machine, controls the allocation of virtual machines to hosts and assigns IP addresses to virtual machine instances. It runs in the Apache Tomcat container and requires a MySQL server as the database.

2. The Cloud infrastructure server: This manages, (i) Regions: One or more cross-region proximate zone/s, (ii) Zones: A single data center, (iii) Pods: A rack or row of racks that include the Layer-2 switch, (iv) Clusters: One or more homogeneous hosts, (v) Host: A single compute node within a cluster, (vi) Primary storage: Storage in a single cluster, (vii) Secondary storage: A zone-wide resource to store disk templates, ISO images and snapshots.

10.6.1   Using the OpenStack Dashboard for Administration

As a Cloud administrative user, we can utilize the OpenStack dashboard to create and manage projects, clients, images etc. Regularly, the strategy arrangement permits administrator clients just to set shares and make and manage services. The dashboard provides an Admin tab with a System Panel and

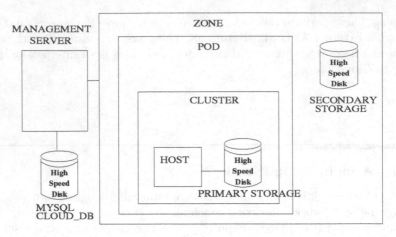

FIGURE 10.4
Apache CloudStack Architecture

Identity Panel. These interfaces give us access to framework data and utilization as settings for arranging what end clients can do.

10.6.2   Command-Line Tools

The combination of the OpenStack command-line interface (CLI) tools and the OpenStack dashboard for administration is recommended for users. Some users with a background in other Cloud technologies may be using the EC2 Compatibility API, which uses naming conventions somewhat different from the native API.

We need to install the command-line clients from the Python Package Index (PyPI) instead of from the distribution packages. The clients are under heavy development, and it is very likely at any given time that the version of the packages distributed by your operating-system vendor are out of date.

The pip utility is used to manage package installation from the PyPI archive and is available in the python-pip package in most Linux distributions. Each OpenStack project has its own client, so depending on which services your site runs, we need to install some or all of the packages: python-novaclient (nova CLI), python-glanceclient (glance CLI), python-keystoneclient (keystone CLI), python-cinderclient (cinder CLI) , python-swiftclient (swift CLI), and python-neutronclient (neutron CLI).

Installing the Tools:
To install (or upgrade) a package from the PyPI archive with pip, as root:

pip install [–upgrade] <package-name>

To remove the package:
pip uninstall <package-name>

If you need even newer versions of the clients, pip can install directly from the upstream git repository using the -e flag. You must specify a name for the Python egg that is installed. For example:
pip install -e git+https://github.com/openstack/
python-novaclient.git#egg=python-novaclient

If you support the EC2 API on your Cloud, you should also install the euca2ools package or some other EC2 API tool so that you can get the same view your users have. Using EC2 API-based tools is mostly out of the scope of this guide, though we discuss getting credentials for use with it.

Administrative Command-Line Tools

There are also several ∗– manage command-line tools, namely nova-manage, glance-manage, keystone-manage and cinder-manage. These are installed with the project's services on the Cloud controller and do not need to be installed separately.

Unlike the CLI tools mentioned above, the ∗– manage tools must be run from the Cloud controller, as root, because they need read access to the config files such as /etc/nova/nova.conf and to make queries directly against the database rather than against the OpenStack API endpoints.

10.6.3   Create an OpenStack Development Environment

To create a development environment, you can use DevStack. DevStack is essentially a collection of shell scripts and configuration files that builds an OpenStack development environment for you.

To run DevStack for the stable Havana branch on an instance in your OpenStack Cloud:

Boot an instance from the dashboard or the nova command-line interface (CLI) with the following parameters:

- Name: devstack-havana

- Image: Ubuntu 12.04 LTS

- Memory Size: 4 GB RAM

- Disk Size: minimum 5 GB

If you are using the nova client, specify –flavor 3 for the nova boot command to get adequate memory and disk sizes. Later, log in and set up DevStack. Here's an example of the commands you can use to set up DevStack on a virtual machine:

a. Log in to the instance:

$ ssh username@my.instance.ip.address

b. Update the virtual machine's operating system:

apt-get -y update

c. Install git:

apt-get -y install git

d. Clone the stable/havana branch of the devstack repository:

$ git clone https://github.com/openstack-dev/devstack.git -b stable/havana devstack/

e. Change to the devstack repository:

$ cd devstack

10.7   AWS Cloud Development Kit (AWS CDK)

The AWS Cloud Development Kit (AWS CDK) is an open source software development framework to model and provision our Cloud application resources using familiar programming languages. Provisioning Cloud applications can be a challenging process that requires us to perform manual actions, write custom scripts, maintain templates, or learn domain-specific languages. AWS CDK uses the familiarity and expressive power of programming languages for modeling our applications. It provides us with high-level components that pre-configure Cloud resources with proven defaults, so that we can build Cloud applications.

Developing with the AWS CDK: Code snippets and longer examples are available in the AWS CDK's supported programming languages: TypeScript, JavaScript, Python, Java, and C#. The AWS CDK Toolkit is a command line tool for interacting with CDK apps. It enables developers to synthesize artifacts such as AWS CloudFormation templates, and deploy stacks for development of AWS accounts.

The AWS Construct Library includes a module for each AWS service with constructs that offer rich APIs that encapsulate the details of how to create resources for an Amazon or AWS service. The aim of the AWS Construct Library is to reduce the complexity and glue logic required when integrating various AWS services to achieve your goals on AWS.

Advantages of the AWS CDK include the following:

- Use logic (if statements, for-loops, etc) when defining the infrastructure

- Use object-oriented techniques to create a model of the system

- Define high level abstractions, share them, and publish them to the project team, company, or community

- Organize the project into logical modules

- Share and reuse the infrastructure as a library

- Testing infrastructure code using industry-standard protocols

- Using the existing code review workflow

- Code completion within the Integrated Development Environment (IDE)

10.8   Windows Azure SDK

The Azure SDKs are collections of libraries built to make it easier to use Azure services from your language of choice. These libraries are designed to be consistent, approachable, diagnosable, dependable and idiomatic.

The benefits offered by the Azure are unique and business-centered. To mention a few, they are given as follows.

1. Apps management: IaaS benefits the organization to build, deploy, and manage the apps in a quick and easier way. An organization can launch the website or create a web app and maintain the infrastructure by customizing the cloud software.

2. Flexibility: Azure delivers an appreciable level of flexibility giving you the option to have functionality as required. You can pay as you consume, switch to Azure, accommodate the business fluctuations and, etc. Thus, there is no worry about the infrastructure all the time.

3. Agility: Azure is fast in terms of deployment, operation, and scalability. This gives a competitive advantage for the companies to adopt Azure. Being the most up to date cloud technology, the infrastructure and applications could be made agile.

4. Compliance: The data stored is in compliance with the regulations which is very much helpful especially for the legal or the finance sectors. It is entirely built around the security and privacy demands that any of the business would readily take the venture.

5. Storage: It is known that Azure has several data centers and delivery points. It facilitates for faster content delivery, optimal user experience, store any data and also able to share the data across the virtual machines as required at a reliable and faster rate.

6. Security: The data on Azure is protected by the spy-movie environment. The data centers have two-tier authentication, proxy card access readers, hand geometry biometric readers, global incident response team. Thus hacking is reduced to a greater extent.

7. Analytics Support: Azure has built-in support to analyze data and derive insights which help in managed SQL services, Machine Learning, Stream and Cortana Analytics. Thus, your business decisions are made smarter leading to new opportunities.

Installation of Windows Azure SDK

1. Go to official site of Windows Azure at link http://www.windowsazure.com/en-us/

2. In the bottom of page, check for "Develop" option. Click on "Show Me More"

3. Now, we need to choose the language we want to work with. We can develop application in any of the language given in option and deploy it on the Microsoft managed datacenters.

4. To start developing using .NET, click on .NET.

5. We will be navigated to Home Page of .NET Developer Center. We get all the resources related to .NET development on Windows Azure.

6. Now click on Install to install Windows Azure SDK.

Summary

Deploying applications in the Cloud can be significantly different than doing so in your own data center. A number of steps go into deploying an application in the Cloud, from build to package to release and there can be a lot of overhead in doing that manually. More and more software development teams are embracing the practice of continuous integration and continuous deployment (CI/CD) as they seek to automate deploying new code into production. This chapter dealt with some of the important tools available that enable a developer to build and deploy an application without having to download anything to their desktop. This requires an on-demand development tool that sits on top of the Cloud and provides Platform-as-a-Service (PaaS).

Objective type questions

1. Which of the following organization supports the development of standards for Cloud computing ?

(a) OPC   (b) OCP

(c) OCC   (d) All of the above

2. Which of the following is a payment processing system by Google ?

(a) Paytm   (b) Checkout

(c) Code   (d) All of the above

3. Which of the following google product sends you periodic email alerts based on your search term ?

(a) Alerts   (b) Blogger

(c) Calendar   (d) All of the above

4. Which of the following service creates an application hosting environment ?

(a) EBS   (b) Azure AppFabric

(c) ESW   (d) All of the above

5. Database marketplace based on SQL Azure Database is code-named _____

(a) Akamai   (b) Dallas

(c) Denali   (d) None of the above

6. Google _____ is the most widely used Web traffic analysis tool on the Internet.

(a) BigAnalytics   (b) Analytics

(c) Biglytics   (d) All of the above

7. Which of the following service that provides offline access to online data ?

(a) Gears   (b) Blogger

(c) Offline   (d) All of the above

8. Which of the following is a key Cloud service attribute ?

(a) Abstraction   (b) Infrastructurer

(c) User Interface   (d) All of the above

9. Which of the following factors might offset the cost of offline access in hybrid application ?

(a) Scalability   (b) Costs

(c) User Interface   (d) Ubiquitous access

10. An application that needed _____ storage alone might not benefit from a Cloud deployment at all.

(a) Online   (b) Offline

(c) Virtual   (d) All of the above

11. Which of the following provider rely on virtual machine technology to deliver servers?

(a) CaaS   (b) AaaS

(c) PaaS   (d) IaaS

12. Point out the correct statement :

(a) Virtual machines are containers that are assigned specific resources

(b) Applications need not be mindful of how they use Cloud resources

(c) When a developer creates an application that uses a Cloud service, the developer cannot attach to the appropriate service

(d) All of the above

13. Which of the following component is called hypervisor ?

(a) VMM

(b) VMc

(c) VMM

(d) All of the above

14. Applications such as a Web server or database server that can run on a virtual machine image are referred to as:

(a) Virtual server

(b) Virtual appliances

(c) Machine imaging

(d) All of the above

15. Point out the wrong statement :

(a) Platforms represent nearly the full Cloud software stack

(b) A platform in the Cloud is a software layer that is used to create higher levels of service

(c) Platforms often come replete with tools and utilities to aid in application design and deployment

(d) None of the above

Objective type questions & answer

1:c 2:b 3:a 4:b 5:b 6:b 7:a 8:b 9:d 10:b 11:d 12:a 13:c 14:b 15:d

Review questions

1. How to use tools such as Xen and KVM to compromised VMs during live migration in the linux Cloud environment?
2. In an IaaS (Xen-based) Cloud, how much monitoring information can the Infrastructure Provider (IP) get about how each VM on a host consumes resources?
3. What is the best way to set up an experiment environment for Cloud performance testing for live VM migration?
4. How can one get a pre-copy dataset at the time of live migration?

Critical thinking questions

1. Compare the three Cloud computing delivery models, SaaS, PaaS and IaaS, from the point of view of the application developers and users. Discuss the security and the reliability of each one of them. Analyze the differences between the PaaS and the IaaS.

2. An organization debating whether to install a private Cloud or to use a public Cloud, e.g., the AWS, for its computational and storage needs, asks your advice. What information will you require to base your recommendation on, and how will you use each one of the following items: (a) the description of the algorithms and the type of the applications the organization will run; (b) the system software used by these applications; (c) the resources needed by each application; (d) the size of the user population; (e) the relative experience of the user population; and (d) the costs involved.

3. What is in your opinion the critical step in the development of a systematic approach to all-or-nothing atomicity? What does a systematic approach means? What are the advantages of a systematic versus an ad-hoc approach to atomicity? The support for atomicity affects the complexity of a system. Explain how the support for atomicity requires new functions/mechanisms and how these new functions increase the system complexity. At the same time, atomicity could simplify the description of a system; discuss how it accomplishes this. The support for atomicity is critical for system features which lead to increased performance and functionality such as: virtual memory, processor virtualization, system calls, and user-provided exception handlers. Analyze how atomicity is used in each one of these cases.

4. Use the start-time fair queuing (SFQ) scheduling algorithm to compute the virtual startup and the virtual finish time for two threads a and b with weights $w\_a = 1$ and $w\_b = 5$ when the time quantum is $q = 15$ and thread b blocks at time $t = 24$ and wakes up at time $t = 60$. Plot the virtual time of the scheduler function of the real time.

Bibliography

[1] Kenel Virtual Machine, *Available at: http://www.linux-kvm.org/page/Main_Page*, Accessed on March 19, 2019.

[2] Cordeiro T., Damalio D., Pereira N., "Open source cloucomputing platforms", *Proceedings of 2010 9th IEEE International Conferencon on Grid and Cooperative Computing (GCC)*, pp. 366-371.

[3] Diez O., Silva, A., "GovCloud: Using Cloud Computing iPublic Organizations", *IEEE Technology and Society Magazine*, Vol. 32, No. 1, 2016, pp. 66-72.

[4] Fremdt S., Beck R., Weber S., "Does Cloucomputing matter? An analysis of the Cloud Model Software-as-a-Service and its impact on operational agility", *Proceedings of 46th IEEE Hawaii International Conference on ISystem Sciences (HICSS)*, 2013, pp. 1025-1034.

[5] J. J. Peng, X. J. Zhang, Z. Lei, B. F. Zhang, W. Zhang, and Q. Li, "Comparison of Several Cloud Computing Platforms", *Proceedings of the Second IEEE International Symposium on Information Science and Engineering*, IEEE Computer Society, Washington, DC, USA, 2009, pp. 23-27.

[6] Haji A., Ben Letaifa A., Tabbane S., "CloudComputing: Several Cloud-oriented Solutions", *Proceedings of Fourth International Conference on Advanced Engineering Computing and Applications in Sciences*, 2010, pp. 137-141.

[7] Garlick, G., "Improving Resilience with Community ClouComputing", *Proceedings of sixth International Conference on Availability, Reliability and Security (ARES)*, 2011, pp. 650- 655.

[8] Ameen M. N., Sanjay H. A., Patel Y., " High Performance Computing Applications", *Proceedings of second IEEE International Conference on Parallel Distributed and Grid Computing (PDGC)*, Bangalore, India, 2010, pp. 262-267.

11

Cloud Security

After reading this chapter, you will be able to

- Assess data security in the Cloud

- Describe security management in the Cloud

- Compare and contrast IaaS, PaaS, SaaS security in the Cloud deployments.

Cloud computing security or, in other words, Cloud security alludes to a wide arrangement of approaches, advances and controls conveyed to ensure information, applications and the related framework of Cloud computing. It is a sub-domain of computer security, network security, and, all the more extensively, data security. Cloud computing storage capability provides clients with abilities to store and procedure their information in data centers. Organizations utilize the Cloud in a wide range of service models (for example, SaaS, PaaS and IaaS) and deployment models (private, hybrid, public and community).

Security concerns related with Cloud computing fall into two categories: security issues faced by Cloud providers (organizations providing software, platform-, or infrastructure-as-a-service via the Cloud) and security issues faced by their clients (organizations or associations who have applications or stored information on the Cloud). The responsibilities are shared among them. The providers must guarantee that their services are secure and that their customer's information and applications are protected, while the client must take measures to support security measures (such as usage of strong passwords).

Preliminaries

The following are some of terms and terminologies given here for easier understanding in the remaining part of this chapter.

Confidentiality is the prevention of the intentional or unintentional unauthorized disclosure of contents. Loss of confidentiality can occur in many ways. For example, loss of confidentiality can occur through the intentional release of private company information or through a misapplication of network rights. Consider the confidentiality of data stored by a distributed storage system. If stored resources are served to any requesting entity, the confidentiality of data intended to be accessible by a specific subset of users cannot be guaranteed.

Integrity is the assurance that the message sent is the message received and that the message is not purposefully or unexpectedly changed. Loss of trustworthiness can happen through a purpose to change data. Consider the trustworthiness of information stored in a Cloud. On the off chance that both system components and consumers acknowledge control messages and information moved aimlessly, an attacker might impact either the server or the customer to acknowledge invalid information by modifying the stream of information sent.

Availabilty alludes to the components that create reliability and stability in systems and frameworks. It guarantees that availability is open when required, permitting approved clients to get the system or frameworks. The performance or scalability remains a challenge inspite of the fact that secure access control, storage, and transmission may permit an appropriate storage framework to protect the secrecy of information.

With the appropriation of open Cloud benefits, a huge part of network, system, framework, applications and information will move under third-party control. The Cloud administrations conveyance model will make islands (Clouds) of virtual perimeters just as a security model with obligations shared between the client and the Cloud service provider (CSP).

11.1   Data Security

Data security becomes very significant for data that is communicated from one device to another. It is refereed to as data-in-transit. This is an information security concept that is used to identify data that requires encryption. The following are common examples of data in transit.

Public Networks: Transfer of data over public networks such as the Internet. For example, an app. on a mobile phone connects to a banking service to request a transaction. Such data is typically encrypted using protocols such as Hyper Text Transfer Protocol (HTTP).

Private Networks: Transfer of data over a secured private network such as an local area network set up for an office location. The principle of defense in depth suggests that systems and applications should not trust that a network has been properly secured. As such, it is common to apply the same level of encryption for private networks as public networks.

Local Devices: Transfer of data between local devices such as computers, data storage devices and peripherals. Such data could be intercepted by the medium that is used for the transfer. For example, a Universal Serial Bus (USB) cord that connects a data storage device and a computer could potentially record or transmit the data.

Data-in-transit is defined into two categories: (i) information that flows over the public or untrusted network such as the Internet, and (ii) data that flows in the confines of a private network such as a corporate or enterprise Local Area Network (LAN). With regard to data-in-transit, the primary risk is in not using a vetted encryption algorithm. Although this is obvious to information security professionals, it is not common for others to understand this requirement when using a public Cloud, regardless of whether it is IaaS, PaaS, or SaaS.

In addition to the security of data, customers should also be concerned about what data the provider collects and how the Cloud Service Provider (CSP) protects that data. Specifically with regard to customer's data, what metadata does the provider have about data, how is it secured, and what access does the customers have to that metadata? As the volume of data with a particular provider increases, so does the value of that metadata.

There are numerous security issues for Cloud computing as it encompasses many technologies including networks, databases, operating systems, virtualization, resource scheduling, transaction management, load balancing, concurrency control and memory management. For example, the network that interconnects the systems in a Cloud has to be secure and mapping the virtual machines to the physical machines has to be carried out securely. Data security involves encrypting the data as well as ensuring that appropriate policies are enforced for data sharing. Some of the significant security concerns in a Cloud computing environment are discussed below.

Data Transmission: It is the process of sending digital or analog data over a communication medium to one or more computing, network, communication or electronic devices. It enables the transfer and communication of devices in a point-to-point, point-to-multipoint and multipoint-to-multipoint environment.

Virtualization Security: Virtualization security is the collective measures, procedures and processes that ensure the protection of a virtualization infrastructure/environment. It addresses the security issues faced by the components of a virtualization environment and methods through which it can be mitigated or prevented.

Network Security: It is an organization's strategy and provisions for ensuring the security of its assets and all network traffic. Network security is manifested in an implementation of security hardware and software.

Data Integrity: It is the aspect of information technology (IT) that deals with the ability an organization or individual has to determine what data in a computer system can be shared with third parties.

Data Privacy: It is a fundamental component of information security. In its broadest use, "data integrity" refers to the accuracy and consistency of data stored in a database, data warehouse, data mart or other construct.

Data Location: Cloud storage is a model of computer data storage in which the digital data is stored in logical pools. The physical storage spans multiple servers (sometimes in multiple locations), and the physical environment is typically owned and managed by a hosting company.

Data Availability: It is a term used by some computer storage manufacturers and storage service providers (SSPs) to describe products and services that ensure that data continues to be available at a required level of performance in situations ranging from normal through "disastrous".

In the next section, we discuss encryption techniques in Cloud.

11.2   Encryption Techniques in Cloud

Cloud encryption is the transformation of a Cloud service customer's data into ciphertext. Cloud encryption is a service offered by Cloud storage providers whereby data, or text, is transformed using encryption algorithms and is then placed on a storage Cloud. Some of the important encryption techniques are discussed below.

1. Attribute-based encryption (ABE): It is a type of public-key encryption in which the secret key of a user and the ciphertext are dependent upon attributes (e.g. the country in which he lives, or the kind of subscription he has). In such a system, the decryption of a ciphertext is possible only if the set of attributes of the user key matches the attributes of the ciphertext. It is further categorized as follows:

   - Ciphertext-policy ABE (CP-ABE): In the CP-ABE, the encryptor controls access strategy. The main idea of CP-ABE is focused on the design of the access structure.

- Key-policy ABE (KP-ABE): In the KP-ABE, attribute sets are used to describe the encrypted texts and the private keys are associated to specified policy that users will have.

2. Fully homomorphic encryption (FHE): It allows computations on encrypted data, and also allows computing sum and product for the encrypted data without decryption.

3. Searchable encryption (SE): It is a cryptographic system which offer secure search functions over encrypted data. SE schemes can be classified into two categories: SE based on secret-key (or symmetric-key) cryptography, and SE based on public-key cryptography. In order to improve search efficiency, symmetric-key SE generally builds keyword indexes to answer user queries.

11.2.1 Decentralized Key-Policy Attribute-based Encryption

In a decentralized scheme, it is not necessary to maintain a fixed number of attribute authorities. Any attribute authority can join and/or leave the system at any time without rebooting the system. First of all, we will look at the sub-modules required and security game followed by the privacy-preserving decentralized KP-ABE.

Sub-modules: The module consists of five sub-modules namely global setup, authority setup, key issuing, encryption and decryption. Let us briefly explain the functionalities of each sub-modules.

Global Setup: This module takes a security parameter as input and output system parameters. These system parameters can be used by authorities who join the system.

Authority Setup: Each attribute authority uses the system parameters obtained from the global setup to generate public and private keys for the attributes it maintains.

Key Issuing: User and attribute authority interact via anonymous key issuing protocol in order to determine a set of attributes belongs to the user. Then attribute authority generates decryption credentials for those attributes and send them to the user.

Encryption: The encryption algorithm takes a set of attributes maintained by attribute authority and the data as input. Then it outputs the ciphertext of the data.

Decryption: The decryption algorithm takes the decryption credentials received from attribute authorities and the ciphertext as input. The decryption will be successful if and only if the user attributes satisfy the access structure.

11.2.2   Security Game

In order to avoid the security vulnerabilities, ABE schemes is proven to be secure against the selective identity (ID) model. In the selective ID model, the adversary provides the identities of the attribute authorities (challenge identities) he wishes to challenge the challenger with. Then challenger (i.e., the system) generates necessary parameters corresponding to the challenge identities and sends them to the adversary. Then the adversary is allowed to make secret queries about the challenge identities. If the adversary cannot decrypt the encrypted message at the end with non-negligible advantage then the scheme discussed here is secure against the selective ID model. Formally, this is represented by a game between the adversary and the challenger. Adversary sends a list of attribute sets and attribute authorities including corrupted authorities to the challenger. Now the challenger generates public and private keys corresponding to the attributes and authorities provided by the adversary. Challenger provides public and private keys corresponding to the corrupted authorities to the adversary while only public keys corresponding to the remaining authorities to the adversary.

Secret Key Queries

• The adversary is allowed to make any number of secret key queries as he wants against the attribute authorities.

• However, the only requirement is that for each user, there must be at least one non corrupted attribute authority from which the adversary can get insufficient number of secret keys.

Challenge

• The adversary sends two messages m_0 and m_1 to the challenger in plain domain.

• Now the challenger randomly chooses one of the messages and encrypt it and send the ciphertext to the adversary.

More Secret Key Queries

• The adversary is allowed to make more secret key queries as long as he satisfies the requirement given earlier.

Guess

• Now the adversary guesses which message was encrypted by the challenger.

• The adversary is said to be successful if he guesses the correct message with probability $\frac{1}{2} + \epsilon$ whereby ϵ is a non-negligible function.

11.2.3 Fully Homomorphic Encryption (FHE)

In FHE, the system involves multiple parties playing three key roles in the secure computation, the *users* who outsource the computation to a remote party (Cloud), with their secret data protected from unauthorized interrogation; the *homomorphic computation (HC) nodes* provided by the Cloud services using unprotected CPU, GPU or FPGA-based instances.

The data flow among participating parties with the FHE system is simple. The users first verify the configuration of the Cloud through remote attestation, and establish the shared secret key with the bootstrapping nodes. Afterwards, the users provision their encryption parameters as well as the secret and public keys to the bootstrapping nodes through the established secret channel. The user's data encrypted under the homomorphic secret key is sent to the HC nodes to perform homomorphic computations. If the computation requires private data from multiple users, each user sends the data encrypted using their own key to the HC nodes.

When bootstrapping is needed in the homomorphic computation, the current intermediate ciphertext is sent from the HC nodes to the bootstrapping nodes. The bootstrapping nodes, running inside a secure enclave, first decrypt the ciphertext, then re-encrypt it using the secret key and send the refreshed ciphertext back to the HC nodes. It removes the noise in the ciphertext, and thus enables further homomorphic computation by the HC nodes. After the whole homomorphic computation is completed, the ciphertext is sent from the HC node back to the users. The users decrypt the ciphertext to retrieve the computation result.

Homomorphic encryption can be used for privacy-preserving outsourced storage and computation. This allows data to be encrypted and out-sourced to commercial Cloud environments for processing, all while encrypted. In highly regulated industries, such as health care, homomorphic encryption can be used to enable new services by removing privacy barriers inhibiting data sharing. For example, predictive analytics in health care can be hard to apply due to medical data privacy concerns, but if the predictive analytics service provider can operate on encrypted data instead, these privacy concerns are diminished.

11.2.4 Searchable Encryption

Searchable encryption works by exposing segment of information of a context that can be used to identify a context. For example, if there is post about electoral debate in BBC, transcript of the debate can be easily identified by "electoral debate", "BBC" and such keywords They along with date may be sufficient to find that specific manuscript among thousand other.

Searchable encryption enables identification of an encrypted content based on some segment of information available about the content with out exposing the content itself. In this example, those key words would be able to find the transcript of electoral debate but won't be enough to get the actual content. So the content owner would encrypt the data with a private key and share the keyword encrypted with the public key of the search provider, or a shared secret key that is shared with the provider as well. Now once the search query comes, the provider would open the envelop of the keyword collection and matches the search. Once the match is found it would establish the connection between the content owner and the requester. Then the requester and content owner can negotiate the terms of accessing the private content with out the presence of the search provider.

Let's see another example of searchable encryption. Suppose Alice wants to forward all the mails marked "Urgent" to John while she is away on her vacation. Now Bob sends an encrypted email to Alice encrypted with Alice's public Key. But the gateway provider has no way of knowing whether to route the mail to John if is it marked "Urgent." Then John extracts the keyword urgent, performs searchable encryption or Public Key Encryption with keyword search on the extracted keyword and send the data to gateway. The gateway can now decrypt the keyword list and route the message to John.

11.2.5   Web Processing Service (WPS)

Towards implementing a Web Processing Services (WPS) on a Cloud, the Open Geospatial Consortium (OGC) WPS defines a standardized interface that facilitates the publishing of geospatial processes so that geoprocessing services can be delivered over the Internet (OGC 2007). Geospatial processing refers to the manipulation of geographic information, ranging from simple feature overlays and geocoding to raster processing and advanced climate modelling.

There are many advantages of deploying a geoprocessing service in a Cloud. For example, a compute intensive WPS, such as one that does climate modelling, could benefit from the scalability of the resources in a Cloud or a WPS that suddenly becomes popular (e.g. one that does polygon intersections) could benefit from the user scalability provided by the Cloud. While geoprocessing services probably stand to benefit most from processing (resource) scalability, implementations of OGC's Web Feature Service (WFS) stand to benefit from the scalability of user requests that a Cloud provides. A WPS can be deployed in any of the three service models of a Cloud, namely IaaS, PaaS, or SaaS.

In an IaaS Cloud, the user has full control over resources, but virtualization happens at a low level, and the user therefore has the responsibility to address many security risks. Deploying geoprocessing on an IaaS Cloud makes sense if an organisation needs to deploy a variety of WPSs and has the required human resources to administer the IaaS Cloud.

In a PaaS Cloud, virtualization happens at a higher level of abstraction, and therefore, some of the development challenges of building scalable applications are handled by the Cloud platform. In a WPS, input data are supplied either embedded in the execute request or referenced as a web accessible resource. Both ways are possible in a PaaS Cloud because they do not rely on accessing files on a local file system. However, the WPS source code cannot create any temporary files that could be required during processing. Instead, such data have to be stored temporarily in Cloud database tables.

Google App Engine (GAE) provides authentication though its web portal and single sign-on authentication for first-time registration. Single sign-on allows a generated password to be used only once. Upon sign up to use GAE and request for a domain name, a password is generated and sent to the user's cellular phone. This generated password can then be used to activate the user's account, after which the password cannot be used again.

All of the Clouds allow Secure Sockets Layer (SSL) encryption. SSL is one of the most widely used communication protocols on the Internet. It provides encrypted communication between a client and server, as well as optional client and server authentication. While SSL provides network communication encryption, data encryption is provided by programming libraries and database management systems.

In the next section, we discuss Infrastructure Security.

11.3   Infrastructure Security

Information security, sometimes shortened to InfoSec, is the practice of preventing unauthorized access, use, disclosure, disruption, modification, inspection, recording or destruction of information. The information or data may take any form, e.g. electronic or physical. Information security's primary focus is the balanced protection of the Confidentiality, Integrity and Availability of data (also known as the CIA triad) while maintaining a focus on efficient policy implementation, all without hampering organization productivity. This is largely achieved through a multi-step risk management process that identifies assets, threat sources, vulnerabilities, potential impacts, and possible controls, followed by assessment of the effectiveness of the risk management plan.

To properly understand the threats that Cloud computing presents to the computing infrastructure, it is important to understand communications security techniques to prevent, detect, and correct errors so that integrity, availability, and the confidentiality of transactions over networks may be maintained. This includes the following:

- Communications and network security as it relates to voice, data, multimedia and facsimile transmissions in terms of local area, wide area and remote access networks

- Internet/intranet/extranet in terms of firewalls, routers, gateways and various protocols

11.3.1   Infrastructure Security: The Network Level

When looking at the network level of infrastructure security, it is important to distinguish between public Clouds and private Clouds. With private Clouds, there are no new attacks, vulnerabilities, or changes in risk specific to this topology that information security personnel need to consider. Although the organization's IT architecture may change with the implementation of a private Cloud, the current network topology will probably not change significantly. If we have a private extranet, an intranet that can be partially accessed by authorized outside users, enabling businesses to exchange information over the Internet in a secure way, for practical purposes we probably have the network topology for a private Cloud in place already. The security considerations we have today apply to a private Cloud infrastructure, too. And the security tools we have in place (or should have in place) are also necessary for a private Cloud and operate in the same way.

Figure 11.1 shows the topological similarities between a secure extranet and a private Cloud. Internet, intranet and extranet are the networks of computer. Internet is an open to all network that can be accessed by anyone in the world with an Internet connected device. Intranet is the computer network for groups containing small members and can be accessed by these members only. These network are used in business fields as one member has to share files and other docs to other members within a group. It is also used for networking within building or home.

Extranet is computer network used by people of particular network with a login identity and password. These are sometimes used by two or more departments to communicate with each other or by some companies to handle account for particular customers. There are three significant risk factors in this usecase, given as follows.

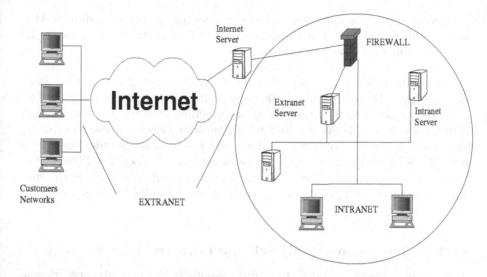

FIGURE 11.1
Similarities between a secure extranet and a private Cloud

Ensuring the confidentiality and integrity of our organization's data-in-transit to and from our public Cloud provider: Some resources and data previously confined to a private network are now exposed to the Internet, and to a shared public network belonging to a third-party Cloud provider. Although use of HTTPS (instead of HTTP) would have mitigated the integrity risk, users not using HTTPS (but using HTTP) did face an increased risk that their data could have been altered in transit without their knowledge.

Ensuring proper access control (authentication, authorization and auditing) to whatever resources we are using at your public Cloud provider: Since some subset of these resources (or maybe even all of them) is now exposed to the Internet, an organization using a public Cloud faces a significant increase in risk to its data. The ability to audit the operations of your Cloud provider's network (let alone to conduct any realtime monitoring, such as on your own network), even after the fact, is probably non-existent.

Ensuring the availability of the Internet-facing resources in a public Cloud that are being used by our organization, or have been assigned to our organization by public Cloud providers: Reliance on network security has increased because an increased amount of data or an increased number of organizational personnel now depend on externally hosted devices to ensure the availability of Cloud-provided resources.

Consequently, the three risk factors enumerated in the preceding section must be acceptable to an organization. Misconfiguration may still affect the availability of your Cloud-based resources.

According to a study presented to the North American Network Operators Group (NANOG) in February 2006, several hundred such misconfigurations occur per month. Probably the best known example of such a misconfiguration mistake occurred in February 2008 when Pakistan Telecom made an error by announcing a dummy route for YouTube to its own telecommunications partner, PCCW, based in Hong Kong. The intent was to block YouTube within Pakistan because of some supposedly blasphemous videos hosted on the site. The result was that YouTube was globally unavailable for two hours.

11.3.2   Trust Zones (TZ) Scheme Infrastructure Security

The vulnerability of Cloud Computing Systems (CCSs) to advanced persistent threats (APTs) is a significant concern to government and industry. Cloud Service Provider (CSP) personnel with physical access to the CSP datacenter can breach security controls by direct access to physical machines. However, the large number of machines complicates the task. Precision requires the attacker first identify the hardware hosting the target data. The CSP insider cannot make physical contact with every machine in the datacenter, but he has several methods to locate the machine hosting the agency Trust Zones (TZs) VMs.

TZs is defined as a combination of network segmentation and identity and access management (IAM) controls. These define physical, logical, or virtual boundaries around network resources. Cloud TZs can be implemented using physical devices, virtually using virtual firewall and switching applications, or using both physical and virtual appliances. The first is to enumerate all of the Agency's live VMs. CSP management servers hold such data. A CSP sys-admin will have access to this data.

The list of agency VMs might be very large. The attacker must reduce the list so he can visit each machine. Information that can help narrow the list is configuration data. This includes security group and other tenant created configurations that reveal the topology of the tenants resources in the Cloud, specific services that can be assumed based on specific ports that are open, identity and access management data that shows which tenant users have CSP accounts and what they are allowed to do with specific resources, tenant naming conventions for their machine images or instances, and relative memory, disk, CPU and I/O sizing of various instances. These steps are likely to identify, for example, large machines, which host web server, file share, or database processes, and those that have limited access.

Once the CSP insider knows which machines to visit, he must map these to the datacenter layout. Datacenters that are segmented or compartmentalized, or keep access to physical and logical maps separate, will complicate the task. In contrast, datacenters that have a single point of entry will make this attack easier.

The malware injected via physical access provides a breach point for the attacker. The breach itself can provide network access and elevated privileges on the HV hosting the target VM. By beaconing to known attacker control nodes, the attacker can establish a link to control the execution of the rest of the attack.

11.4 PaaS Application Security

PaaS vendors broadly fall into the following two major categories, i.e., software vendors (e.g., Bungee, Etelos, GigaSpaces, Eucalyptus) and CSPs (e.g., Google App Engine, Salesforce.com's Force.com, Microsoft Azure, Intuit QuickBase).

Organizations evaluating a private Cloud may utilize PaaS software to build a solution for internal consumption. Currently, no major public Clouds are known to be using commercial off-the-shelf or open source PaaS software such as Eucalyptus (Eucalyptus does offer a limited experimental pilot Cloud for developers at Eucalyptus.com). Therefore, given the nascent stage of PaaS deployment, it is recommended that organizations evaluating PaaS software perform a risk assessment and apply the software security standard similar to acquiring any enterprise software.

Generally speaking, PaaS CSPs (e.g., Google, Microsoft, and Force.com) are responsible for securing the platform software stack that includes the run-time engine that runs the customer applications. Since PaaS applications may use third-party applications, components, or web services, the third-party application provider may be responsible for securing their services. Hence, customers should understand the dependency of their application on all services and assess risks pertaining to third-party service providers.

Until now, CSPs have been reluctant to share information pertaining to platform security using the argument that such security information could provide an advantage for hackers. However, enterprise customers should demand transparency from CSPs and seek information necessary to perform risk assessment and ongoing security management.

PaaS Access Control

Generally speaking, access control management in PaaS is a broad function that encompasses access requirements for your users and system administrators (privileged users) who access network, system, and application resources. The access control management functions should address questions such as assignment of entitlements to users, assignment of entitlements based on the user's job functions and responsibilities, authentication method and strength required prior to granting access to the resource, auditing and reporting to verify entitlement assignments. The aforementioned aspects of the access control domain should be addressed by organization's access policies and standards and aligned with the user's roles and responsibilities, including end users and privileged system administrators.

Hardening the Host OS for PaaS Security

Vulnerabilities inherent in the operating system of the host computer can flow upward into the virtual machine operating system. While a compromise on the VM OS would hopefully only compromise the guest domain, a compromise of the underlying host OS would give an intruder access to all services on all virtual machines hosted by the machine. Therefore, best practice hardening techniques must be implemented to maintain the security posture of the underlying technology. Some of these techniques include the following:

- Use strong passwords, such as lengthy, hard to guess passwords with letters, numbers, and symbol combinations, and change them often.

- Disable unneeded services or programs, especially networked services.

- Require full authentication for access control.

- The host should be individually firewalled.

- Patch and update the host regularly, after testing on a nonproduction unit.

Single Sign-On (SSO) Scheme for PaaS Security

Single sign-on (SSO) addresses the cumbersome situation of logging on multiple times to access different resources. When users must remember numerous passwords and IDs, they might take shortcuts in creating them that could leave them open to exploitation. In SSO, a user provides one ID and password per work session and is automatically logged on to all the required applications. For SSO security, the passwords should not be stored or transmitted in the clear.

SSO applications can run either on a user's workstation or on authentication servers. The advantages of SSO include having the ability to use stronger passwords, easier administration of changing or deleting the passwords, and less time to access resources. The major disadvantage of many SSO implementations is that once users obtain access to the system through the initial logon, they can freely roam the network resources without any restrictions.

A very popular example of SSO login is Google's implementation for their software products. Once a user is logged in to Gmail, the user automatically gains access to YouTube, Google Drive, Google Photos and other Google products.

11.5   SaaS Application Security

The SaaS model dictates that the provider manages the entire suite of applications delivered to users. Therefore, SaaS providers are largely responsible for securing the applications and components they offer to customers. Customers are usually responsible for operational security functions, including user and access management as supported by the provider. It is a common practice for prospective customers to request information related to the provider's security practices. This information should encompass design, architecture, development, black- and white-box application security testing, and release management.

Some SaaS applications, such as Google Apps, have built-in features that end users can invoke to assign read and write privileges to other users. However, the privilege management features may not be advanced, fine-grained access and could have weaknesses that may not conform to your organization's access control standard. One example that captures this issue is the mechanism that Google Docs employs in handling images embedded in documents, as well as access privileges to older versions of a document.

Evidently, embedded images stored in Google Docs are not protected in the same way that a document is protected with sharing controls. That means, if one has shared a document containing embedded images, the other person will always be able to view those images even after you've stopped sharing the document.

11.5.1 FHE Scheme for SaaS Security

Unlike the partially homomorphic encryption which allow homomorphic computation of some operations on ciphertexts (e.g., additions, multiplications, quadratic functions, etc.), fully homomorphic encryption supports arbitrary computation on ciphertexts. Such a scheme enables the construction of programs for any desirable functionality, which can be run on encrypted inputs to produce an encryption of the result. Since such a program need never decrypts its inputs, it can be run by an untrusted party without revealing its inputs and internal state.

Here is a very simple example of how a homomorphic encryption scheme might work in Cloud computing:

- Business B has a very important data set (VIDS) that consists of the numbers 5 and 10. To encrypt the data set, Business B multiplies each element in the set by 2, creating a new set whose members are 10 and 20.

- Business B sends the encrypted VIDS set to the Cloud for safe storage. A few months later, the government contacts Business X and requests the sum of VIDS elements.

- Business B is very busy, so it asks the Cloud provider to perform the operation. The Cloud provider, who only has access to the encrypted data set, finds the sum of $10 + 20$ and returns the answer 30.

- Business B decrypts the Cloud provider's reply and provides the government with the decrypted answer, 15.

Access Control in SaaS

In a Cloud computing consumption model, where users are accessing Cloud services from any Internet-connected host, network access control will play a diminishing role. The reason is that traditional network-based access controls are focused on protecting resources from unauthorized access based on host-based attributes, which in most cases is inadequate, is not unique across users, and can cause inaccurate accounting. In the Cloud, network access control manifests as Cloud firewall policies enforcing host-based access control at the ingress and egress points of entry to the Cloud and logical grouping of instances within the Cloud. This is usually achieved using policies (rules) using standard Transmission Control Protocol/Internet Protocol (TCP/IP) parameters, including source IP, source port, destination IP, and destination port.

In contrast to network-based access control, user access control should be strongly emphasized in the Cloud, since it can strongly bind a user's identity to the resources in the Cloud and will help with fine granular access control, user accounting, support for compliance, and data protection. User access management controls, including strong authentication, single sign-on (SSO), privilege management, and logging and monitoring of Cloud resources, play a significant role in protecting the confidentiality and integrity of your information in the Cloud. ISO/IEC 27002 has defined six access control objectives that cover end user, privileged user, network, application and information access control.

Similar to SaaS service monitoring, customers who are hosting applications on an IaaS platform should take additional steps to monitor the health of the hosted application. For example, if we are hosting an e-commerce application on an Amazon EC2 virtual Cloud, we should monitor the health of both the e-commerce application and the virtual server instances. The strategies of securing virtual servers are given next.

11.6   Securing Virtual Servers

The simplicity of self-provisioning new virtual servers on an IaaS platform creates a risk that insecure virtual servers will be created. Secure-by-default configuration needs to be ensured by following or exceeding available industry baselines. Securing the virtual server in the Cloud requires strong operational security procedures coupled with automation of procedures. Here are some recommendations by IT organizations.

Use a secure-by-default configuration: VM image must be hardened and used as a standard hardened image for instantiating VMs (the guest OS) in a public Cloud. A best practice for Cloud-based applications is to build custom VM images that have only the capabilities and services necessary to support the application stack. Limiting the capabilities of the underlying application stack not only limits the host's overall attack surface, but also greatly reduces the number of patches needed to keep that application stack secure.

Track the inventory of VM images and OS versions that are prepared for Cloud hosting: The IaaS provider provides some of these VM images. When a virtual image from the IaaS provider is used it should undergo the same level of security verification and hardening for hosts within the enterprise. The best alternative is to provide your own image that conforms to the same security standards as internal trusted hosts. Protect the integrity of the hardened image from unauthorized access.

Safeguard the private keys required to access hosts in the public Cloud: In general, isolation of the decryption keys from the Cloud where the data is hosted is required, unless they are necessary for decryption. If an application requires a key to encrypt and decrypt for continuous data processing, it may not be possible to protect the key since it will be collocated with the application.

Include no authentication credentials in virtualized images except for a key to decrypt the filesystem key: We should not allow password-based authentication for shell access. For administration privileges (sudo *) or role-based access (e.g., Solaris, SELinux), passwords are mandatory.

Run a host firewall and open only the minimum ports necessary to support the services on an instance: The users should run only the required services and turn off the unused services (e.g., turn off FTP, print services, network file services and database services if they are not required).

Enable system auditing and event logging, and log the security events to a dedicated log server: The log server with higher security protection, including accessing controls needs to be isolated.

11.7   Cloud Security Controls

Cloud security architecture is effective only if the correct defensive implementations are in place. An efficient Cloud security architecture should recognize the issues that will arise with security management. The security management addresses these issues with security controls. These controls are put in place to safeguard any weaknesses in the system and reduce the effect of an attack. While there are many types of controls behind a Cloud security architecture, they can usually be found in one of the following categories.

Deterrent controls: These controls are intended to reduce attacks on a Cloud system. Much like a warning sign on a fence or a property, deterrent controls typically reduce the threat level by informing potential attackers that there will be adverse consequences for them if they proceed.

Preventive controls: It strengthens the system against incidents, generally by reducing if not actually eliminating vulnerabilities. Strong authentication of Cloud users, for instance, makes it less likely that unauthorized users can access Cloud systems, and more likely that Cloud users are positively identified.

Detective controls: Detective controls are intended to detect and react appropriately to any incidents that occur. In the event of an attack, a detective control will signal the preventive or corrective controls to address the issue. System and network security monitoring, including intrusion detection and prevention arrangements, are typically employed to detect attacks on Cloud systems and the supporting communications infrastructure.

Corrective controls: It reduces the consequences of an incident, normally by limiting the damage. They come into effect during or after an incident. Restoring system backups in order to rebuild a compromised system is an example of a corrective control.

Summary

Cloud computing is a nascent and rapidly evolving model, with new aspects and capabilities being announced regularly. In this chapter, we provide a comprehensive and timely look at the security issues w.r.t IaaS, Paas and SaaS services. Organizations that rely solely on a Cloud vendor's built-in security potentially expose their organization to unnecessary risk. This is particularly true for the credentials and secrets that proliferate in Cloud environments and automated processes. These secrets are dynamically created and assigned to provision, configure and manage hundreds of thousands of machines and microservices, but many are never secured. If they are compromised, these secrets and credentials can give attackers a crucial jumping-off point to achieve lateral access across networks, data and applications, and ultimately, provide access to an organization's most critical assets.

Keywords

Security management	Data security
IaaS security	PaaS security
SaaS security	

Objective type questions

1. Which of the following service provider provides the least amount of built in security ?

(a) SaaS

(b) PaaS

(c) IaaS

(d) All of the above

2. Point out the correct statement:

(a) Different types of Cloud computing service models provide different levels of security services

(b) Adapting your on-premises systems to a Cloud model requires
that you determine what security mechanisms are required
and mapping those to controls
that exist in your chosen Cloud
service provider

(c) Data should be transferred and
stored in an encrypted format
for security purpose

(d) All of the above

3. Which of the following services that need to be negotiated in Service Level
Agreements ?

(a) Logging

(b) Auditing

(c) Regulatory compliance

(d) All of the above

4. Point out the wrong statement:

(a) You can use proxy and brokerage
services to separate clients from
direct access to shared Cloud
storage

(b) Any distributed application has
a much greater attack surface
than an application that is
closely held on a LAN

(c) Cloud computing don't have vulnerabilities associated with Internet applications

(d) All of the above

5. Which of the following area of Cloud computing is uniquely troublesome ?

(a) Auditing

(b) Data integrity

(c) e-Discovery for legal compliance

(d) All of the above

6. Which of the following is operational domain of CSA ?

(a) Scalability

(b) Portability and interoperability

(c) Flexibility

(d) None of the above

7. Which of the following is considered an essential element in Cloud computing by CSA ?

(a) Multi-tenancy

(b) Identity and access management

(c) Virtualization

(d) All of the above

8. Which of the following is used for Web performance management and load
testing ?

(a) VMware Hyperic   (b) Webmetrics

(c) Univa UD   (d) Tapinsystems

9. Which of the following is application and infrastructure management software for hybrid multi-Clouds ?

(a) VMware Hyperic   (b) Webmetrics

(c) Univa UD   (d) Tapinsystems

10. Which of the following is policy based XML security service by Cisco ?

(a) Application Oriented Manager   (b) Application Oriented Networking

(c) Application Process Networking   (d) All of the above

11. Point out the wrong statement:

(a) SOA eliminates the use of application boundaries, the traditional methods where security is at the application level aren't likely to be effective

(b) An atomic service cannot be decomposed into smaller services that provide a useful function

(c) XML security service may be found in Citrix's NetScaler 9.0

(d) None of the above

12. Which of the following is not a OASIS standard for SOA Security ?

(a) Security Assertion Markup Language

(b) Synchronized Multimedia Integration Language

(c) WS-SecureConversion   (d) All of the above

13. Which of the following provides data authentication and authorization between client and service ?

(a) SAML   (b) WS-SecureConversion

(c) WS-Security   (d) All of the above

14. Point out the wrong statement:

(a) To address SOA security, a set of OASIS standards have been created

(b) WS-SecureConversion attaches a security context token to communications such as SOAP used to transport messages in an SOA enterprise

(c) WS-Trust is an extension of SOA that enforces security by applying tokens such as Kerberos, SAML, or X.509 to messages

(d) None of the above

15. Which of the following is web services protocol for creating and sharing security context ?

(a) WS-Trust   (b) WS-Secure Conversion

(c) WS-Security Policy   (d) All of the above

Objective type questions -answer

1:c 2:d 3:d 4:c 5:d 6:b 7:a 8:b 9:c 10:b 11:d 12:b 13:a 14:c 15:b

Review questions

1. How a Cloud service provider ensures data integrity during transmission (within the Clouds and on the way from/to the Cloud)?
2. How data and applications are separated from one customer data and applications from other clients?
3. How the providers apply laws and regulations that are applicable to Cloud computing?
4. Mention platforms which are used for large scale Cloud computing.
5. What is the difference in Cloud computing and computing for mobiles?
6. How user can gain from utility computing?
7. What are the security aspects provided with Cloud?
8. List out different layers which define Cloud architecture.
9. What does "EUCALYPTUS" stands for?
10. What is the requirement of virtualization platform in implementing Cloud?
11. Mention the name of some large Cloud providers and databases.
12. Explain the difference between Cloud and traditional datacenters?

Critical thinking questions

1. Search the literature for papers proposing a layered architecture for computing Clouds and analyze critically the practicality of each of these approaches.
2. Discuss means to cope with the complexity of computer and communication systems other than modularity, layering and hierarchy.
3. We know how a VMBK (Virtual Machine-Based Rootkit) could pose serious security problems. Give other examples when abstraction of a physical system could be exploited for nefarious activities.
4. Give examples of undesirable behavior of computing and communication system that can be characterized as phase-transitions. Can you imagine a phase-transition in a Cloud?

Bibliography

[1] Chen H, Liu X, Xu H, Wang C, "A Cloud service broker based on dynamic game theory for bilateral SLA negotiation in Cloud environment", *Int J Grid Distrib Comput*, Vol. 9, no. 9, 2018, pp. 251-268.

[2] R. Gellman, "Privacy in the Clouds: Risks to privacy and confidentiality from Cloud computing", *The World Privacy Forum, 2009*, Available at: http://www.worldprivacyforum.org/

[3] Ohlman, B., Eriksson, A., Rembarz, R., "What Networking of Information Can Do for Cloud Computing", *Proceedings of 18th IEEE International Workshops on Enabling Technologies: Infrastructures for Collaborative Enterprises*, Groningen, The Netherlands, June 29-July 1, 2009.

[4] L.J. Zhang and Qun Zhou, "CCOA: Cloud Computing Open Architecture", *IEEE International Conference on Web Services*, pp. 607-616, 2009.

[5] Tim Mather, Subra Kumaraswamy, Shahed Latif, "Cloud Security and Privacy: An Enterprise Perspective on Risks and Compliance", *Book published by O' Reilly Media*, USA, 2009.

[6] Ronald L. Krutz, Russell Dean Vines, "Cloud Security: A Comprehensive Guide to Secure Cloud Computing", *Book published by Wiley Publishing Inc.*, 2010.

[7] K. Vieira, A. Schulter, C. B. Westphall, and C. M. Westphall, "Intrusion detection techniques for Grid and Cloud Computing Environment", *IT Professional Magazine, IEEE Computer Society*, Vol. 12, no. 4, pp. 38-43, 2010.

[8] Marios D. Dikaiakos, Dimitrios Katsaros, Pankaj Mehra, George Pallis, Athena Vakali, "Cloud Computing: Distributed Internet Computing for IT and Scientific Research", *IEEE Internet Computing Journal*, Vol. 13, no. 5, pp. 10-13, September 2009, DOI: 10.1109/MIC.2009.103.

[9] A. Williamson, "Comparing Cloud computing providers", *Cloud Comp. Journal*, Vol. 2, no. 3, pp. 3-5, 2009.

12

Technical and Legal Issues in Cloud Computing

|Learning Objectives|

After reading this chapter, you will be able to

- Assess technical and legal issues in Cloud

- Explore research opportunities in Cloud computing

As we saw before, Cloud computing is Internet based computing. It is located somewhere in the network. It is elastic, cost effective, on demand service, just like electricity on pay-as-per basis. In spite of all these benefits, it raises many technical, legal issues and challenges to Cloud users and Cloud service providers. Cloud computing research addresses the challenges of meeting the requirements of next generation private, public and hybrid Cloud computing architectures.

Likewise, the challenges of permitting applications and development platforms to exploit the advantages of Cloud computing. The exploration on Cloud computing is still at a beginning phase. Many existing issues have not been completely addressed, while new difficulties continue rising up in industry applications. Fig 12.1 shows a diagramatic portrayal of a portion of the difficult issues in Cloud, to be specific accessibility, trust, identity & access, consistency, administration, information security and so on. This section centers around specialized and legitimate security issues from the utilization of Cloud services.

12.1 Technical Issues in Cloud

While companies try to solve Cloud computing issues, the market remains some way off maturity. Gartner recently suggested that Cloud computing as used for service-enabled applications still had seven years until it reached market maturity. Some of the problems it faces are as follows.

Failure of Service: Cyber insurers need to be aware of all the different ways a Cloud provider can fail so that their policy language reflects the risk they

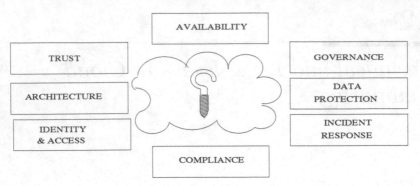

FIGURE 12.1
Challenging issues in Cloud

are intending to take and they can avoid being surprised by non-affirmative, or "silent" cyber risks.

Problem with Internet bandwidth: Bandwidth may not be as prominent concern as energy when it comes to IT and data centers, but it does merit scrutiny. Insufficient bandwidth can strangle the growth of Cloud computing: both at the business and consumer levels.

Data recovery and backup: Data stored at the datacenter is increasing day by day. It leads into huge amount of data storage in Cloud and results into issues such as data loss, data breach etc.

Loss or bankruptcy of service provider: In 2013, Cloud storage provider Nirvanix failed due to cash flow problems, not technical problems. This gives companies a clear view of how exposed they may be, not just because of the service provider's failure, but also because of many issues that have to be addressed to extract themselves from the failure and keep their business operating normally.

Improper Cloud service exit policy: Many Cloud issues, even those that force users to move applications back on premises, will likely be resolved as Cloud services and applications mature. A solid Cloud exit strategy should ensure you do not burn bridges with your provider, for future usage of their services.

Vendor lock-in : Vendor lock-in is a situation in which a customer using a product or service cannot easily make transition to a competitor's product or service. Vendor lock-in is usually the result of proprietary technologies that are incompatible with those of competitors. However, it can also be caused by inefficient processes or contract constraints, among other things.

Multitenancy : Multitenancy is a reference to the mode of operation of software where multiple independent instances of one or multiple applications operate in a shared environment.

Virtualization: Current network defenses are based on physical networks. In the virtualized environment, the network is no longer physical; its configuration can actually change dynamically, which makes network monitoring difficult.

12.1.1 SaaS Issues

In the past couple of years, SaaS has become a staple for many businesses around the world. It presents not only a step forward for the digital age and sharing data, but also as one of the signs that companies are willing to keep up with the latest technologies. While a majority of companies are using them, there are still concerns, risks, and misconceptions regarding their services, discussed as follows.

Data Access Risk: Because users are giving their information and data to a third party, there are concerns about who gets access. It may seem out of their control and fear the potential dissemination, deletion, or corruption of their data by unauthorized people.

Instability: Security and stability are the true pillars that hold up a reliable SaaS software. The services are becoming increasingly popular, which is a double-edged sword. On one hand, it means more options for users and high-quality services because it forces every single provider to keep up with the competition. On the other hand, not everyone will be able to keep up with the growing market. And, in the end, employed provider might shut down because they can no longer compete.

Lack of Transparency: SaaS providers are often secretive and assure their clients that they are better at keeping their data safe than any other. At the very least, they guarantee that they will be capable of securing information and files more proficiently than the customer themselves. However, not all users take their word at face value. There are numerous concerns regarding the provider's lack of transparency on how their entire security protocol is being handled. Unfortunately, this is a matter up for debate.

Identity Theft: SaaS providers always require payment through credit cards that can be done remotely. It is a quick and convenient method, but it does concern some users about the potential risk it implies. There are numerous security protocols placed to prevent problems. It's also severely flawed because this process is still in its infancy. Providers often do not have a better solution for identity management than the company's own firewall.

Uncertainty of Your Data's Location: Most SaaS providers do not disclose where their data centers are, so customers are not aware where it is actually stored. At the same time, they must also be aware of the regulations placed by the Federal Information Security Management Act (FISMA) which states that customers need to keep sensitive data within the country.

Paying Upfront and Long-Term: Financial security is also an issue that may be born out of your agreement to use a SaaS provider. A good majority of them require payment upfront and for long-term. It is more when we are unsure of how long we need their service or if something in their policy will change through time.

No Direct Control Over Your Own Data: Along with concerns that the SaaS provider's servers could shut down for good, there are risks and worries regarding the fact that your data is not really in your control. The good side is that you do not have to configure, manage, maintain, or upgrade the software. The downside of that is that you essentially lose some control over your data.

The Service May Not Keep Up with Modern Security Standards: Plenty of providers boast of their security credentials and prove to their users that they have excellent control over their data and security. However, most will speak of standards that are not up to date, and it does say quite a lot about how mature the service really is.

12.1.2   PaaS Issues

PaaS can significantly impact application performance, availability and flexibility. Anyone involved in administering or customizing a Cloud deployment should have a hand in the decision to deploy PaaS. Here are critical issues to consider.

Open or proprietary?: PaaS is typically designed to work best on each provider's own Cloud platform. The benefit of this is that customers can expect to wring the most value out of the service. The risk is that the customizations or applications developed in one vendor's Cloud environment may not necessarily migrate easily to another vendor. Many customers will be fine with being tied to a single supplier, but if we want to keep our options open, we should ask our Cloud provider for open standards development languages and APIs it supports.

Software compatibility: Most enterprises standardize on a limited set of programming languages, architectural frameworks and databases. We need to be sure that the IaaS vendor we choose supports these technologies. If you are strongly committed to a .NET architecture, for example, then you should select a provider with native .NET support. Similarly, database support is critical to performance and scalability.

Support availability and cost: Using a Cloud provider's proprietary PaaS is like learning a new programming language. It encompasses a learning curve and the need for ongoing support. If the provider's solution is to point you to a library of manuals, that may not be enough for you.

Platform Management: Challenges in delivering middleware capabilities for building, deploying, integrating and managing applications in a multi-tenant, elastic and scalable environments. One of the most important parts of Cloud platforms provide various kind of platform for developers to write applications that run in the Cloud, or use services provided from the Cloud, or both. Different names are used for this kind of platform today, including on-demand platform and PaaS. This new way of supporting applications has great potential.

12.1.3   IaaS Issues

Cloud-computing case studies of customers worldwide should be made available to enterprises to help them clearly understand the areas where Cloud IaaS services benefits can be leveraged. Cloud service providers should invest in developing Cloud labs that enable customers to use Cloud IaaS services on an experimental basis. This is also expected to assist customers in assessing challenges that could arise from integrating enterprise data centers with infrastructure on the Cloud. Some of the challenging research issues in IaaS are discussed here in detail.

Service Level Agreements (SLA): Cloud is administrated by SLAs that allow several instances of one's application to be replicated on multiple servers, if need arises. A big challenge for the Cloud customers is to evaluate SLAs of Cloud vendors. Most vendors create SLAs to make a defensive shield against legal action, while offering minimal assurances to customers. So, there are some important issues, e.g., data protection, outages, and price structures, that need to be taken into account by the customers before signing a contract with a provider.

The specification of SLAs will better reflect the customer's needs if they address the required issues at the right time. Some of the basic questions related to SLA are uptime i.o., are they going to be up 99.90% of the time or 99.99% of the time? And also how does that difference impact the ability to conduct one's business? Is there any SLA associated with backup, archive, or preservation of data ? If the service account becomes inactive, then do they keep user data? If yes, then how long? So it's a important research area in Cloud computing.

Cloud Data Management: Cloud data can be very large (e.g. text-based or scientific applications), unstructured or semi-structured, and typically append-only with rare updates. Since service providers typically do not have access to the physical security system of data centers, they must rely on the infrastructure provider to achieve full data security. However, in a virtualized environment like the Clouds, VMs can dynamically migrate from one location to another; hence directly using remote attestation is not sufficient. In this case, it is critical to build trust mechanisms at every architectural layer of the Cloud.

Data Encryption: Encryption is a key technology for data security. Security can range from simple (easy to manage, low cost and quite frankly, not very secure) all the way to highly secure (very complex, expensive to manage and quite limiting in terms of access). The users and the providers of the Cloud computing solutions have many decisions and options to consider. Once the object arrives at the Cloud, it is decrypted, and stored. Is there an option to encrypt it prior to storing? Do we have to worry about encryption before we upload the file for Cloud computing or do we prefer that the Cloud computing service automatically do it for us? These are options to understand in Cloud computing solution and make decisions based on desired levels of security.

Migration of Virtual Machines: Virtualization can provide significant benefits in Cloud computing by enabling virtual machine migration to balance load across the data center. In addition, virtual machine migration enables robust and highly responsive provisioning in data centers. The major benefits of VM migration is to avoid hotspots; however, this is not straightforward. Currently, detecting workload hotspots and initiating a migration lacks the agility to respond to sudden workload changes.

Interoperability: This is the ability of two or more systems working together to exchange information. Many public Cloud networks are configured as closed systems and are not designed to interact with each other. To overcome this challenge, industry standards must be developed to help Cloud service providers design interoperable platforms and enable data portability. Organizations need to automatically provision services, manage VM instances, and work with both Cloud-based and enterprise-based applications using a single tool set that can function across existing programs and multiple Cloud providers. In this case, there is a need to have Cloud interoperability.

Access Controls: Authentication and identity management is more important than ever. Some concerns, for example: (i) What level of enforcement of password strength and change frequency does the service provider invoke?

What is the recovery methodology for password and account name? (ii) How are passwords delivered to users upon a change? (iii) What about logs and the ability to audit access? This is not all that different from how we secure our internal systems and data, and it works the same way. If we use strong passwords, we should change it frequently, with typical IT security processes.

Energy Resource Management: Significant saving in the energy of a Cloud data center without sacrificing SLAs are an excellent economic incentive for data center operators and would also make a significant contribution to greater environmental sustainability. It has been estimated that the cost of powering and cooling accounts for 53% of the total operational expenditure of data centers. The need arises not only to cut down energy cost in data centers, but also to meet government regulations and environmental standards. Designing energy-efficient data centers has recently received considerable attention.

Multitenancy: Multiple customers accessing the same hardware, application servers, and databases may affect response times and performance for other customers. For application-layer multitenancy specifically, resources are shared at each infrastructure layer and have valid security and performance concerns. For example, multiple service requests accessing resources at the same time increase wait times but not necessarily CPU time, or the number of connections to an HTTP server has been exhausted, and the service must wait until it can use an available connection or in a worst-case scenario, drops the service request.

Server consolidation: The increased resource utilization and reduction in power and cooling requirements achieved by server consolidation are now being expanded into the Cloud. Server consolidation is an effective approach to maximize resource utilization while minimizing energy consumption in a Cloud computing environment.

Reliability and Availability of Service: The challenge of reliability comes into the picture when a Cloud provider delivers on-demand software as a service. The software needs to have a quality factor, such as reliability so that users can access it under any network conditions, such as during slow network connections.

12.2 Performance Challenges in Cloud

What do Amazon EC2, Microsoft Azure, and Google Apps have in common? They are all Cloud computing services, of course. But they share something else in common: each of these Clouds has experienced periods of outages and slowdowns, impacting businesses worldwide that increasingly rely on the Cloud for critical operations. And while there is a great deal of publicity when these prominent public Clouds suffer outages, it's no less damaging to the business when an IT department's private Cloud goes off-line, even if it doesn't make the news.

Moving services to the Cloud promises to deliver increased agility at a lower cost – but there are many risks along the way and greater complexity to manage when you get there. The following are five critical hurdles that you may face when implementing and operating a private Cloud or hybrid Cloud and how you can overcome them.

Will it work?: How can you tell which applications are suitable for Cloud and plan a successful migration?

Performance: If you do not know which physical servers your application is running on, how do you find server-related root causes when performance issues arise?

Chargeback: How do you know how much CPU your application is consuming in order to choose an appropriate chargeback model or verify your bills?

Not aligned with the business: How do you ensure that services are allocated according to business priority?

Over-provisioning: How can you right-size capacity and prevent over-provisioning that undercuts ROI?

12.3   Legal Issues

Cloud computing is bringing amazing advantages and benefits companies. But it also brings some challenges. There are several legal issues that must be taken into consideration when moving into the Cloud. Organizations evaluating Cloud computing should carefully consider a long list of legal issues before taking the plunge, discussed as follows.

Data protection and privacy: Data security has consistently been a major issue in information technology. In the Cloud computing environment, it becomes particularly serious because the data is located in different places even in all the globe. Data security and privacy protection are the two main factors of user's concerns about the Cloud technology.

Lack proper contract and service level agreement (SLA) between Cloud user and provider: SLAs have become more important as organizations move their systems, applications and data to the Cloud. A Cloud SLA ensures that Cloud providers meet certain enterprise-level requirements and provide customers with a clearly defined set of deliverables. The defined level of services should be specific and measureable in each area. This allows the Quality-of-service (QoS) to be benchmarked and, if stipulated by the agreement, rewarded or penalized accordingly.

Conflict of law and multi-jurisdictional issues: In a modern world, where the Internet connects everyone and everything, well beyond borders, is it really reasonable to expect that data in one jurisdiction will only remain in that jurisdiction and not be accessible or discoverable outside that jurisdiction? The implications of hacking, for example, is well known. It needs to be addressed.

Lack of transparency and accountability in data process and data storage: In the past year, Big data has entered the popular consciousness in an amazing way. Many terms that practitioners have used for decades now trip off the tongues of people who literally overheard them for the first time yesterday. In the process of being acquired by new speakers, many big data terms of art have picked up new connotations that may or may not be grounded in reality.

E-discovery and digital investigation: According to the American Bar Association, a mere 10% of all documents created since 1999 are not digitally produced; the vast majority of existing records are now in some type of digital format. Deleting digital documents or failing to retrieve digital records when needed can increase a company's risk of legal liability. To minimize exposure to risk, many companies have employed a data retention process to aid in electronic discovery and computer forensics.

Data deletion, alteration and leak: Cloud and virtualization gives you agility and efficiency to instantly roll out new services and expand your infrastructure. But the lack of physical control, or defined entrance and egress points, bring a whole host of Cloud data security issues: data co-mingling, privileged user abuse, snapshots and backups, data deletion, data leakage, geographic regulatory requirements, Cloud super-admins, and many more.

IPR protection issues: Protection of intellectual property rights from potential infringement has become a herculean task in the Cloud environment. The Cloud, due to its nebulous nature, possesses a number of infringement possibilities. The main concern over the protection of intelloctual property is that they are territorial rights.

Summary

Cloud computing, by its nature, presents fresh challenges to the existing legislation governing the security and privacy of customer data. As Clouds are essentially data centers or server farms used to host and maintain customer data, the customer data is no longer under the complete control of the customer themselves. Traditional licensing agreements and contracts may be legally inadequate, and typically do not provide remedies and legal recourse for specific situations. Also, there is an underlying fear amongst some customers with regards to data security, with the protection and privacy of their data, both from loss and inappropriate distribution, being to the fore. This chapter outlined some of the technical and legal issues associated with Cloud computing and how they may be addressed by ongoing researches in the field of Cloud computing.

Keywords

Research issues	Legal issue
Technical issue	Security issue

Objective type questions

1. Which of the following subject deals with pay-as-you-go usage model?

(a) Accounting Management

(b) Compliance

(c) Data Privacy

(d) All of the above

2. Point out the correct statement:

(a) Except for tightly managed SaaS Cloud providers, the burden of resource management is still in the hands of the user

(b) Cloud computing vendors run very reliable networks

(c) The low barrier to entry cannot be accompanied by a low barrier to provisioning

(d) All of the above

3. _____ captive requires that the Cloud accommodate multiple compliance regimes.

(a) Licensed

(b) Policy-based

(c) Variable

(d) All of the above

4. Security methods such as private encryption, VLANs and firewalls comes under _____ subject area

(a) Accounting Management

(b) Compliance

(c) Data Privacy

(d) All of the above

5. Point out the wrong statement:

(a) Large Cloud providers with geographically dispersed sites worldwide therefore achieve reliability rates that are hard for private systems to achieve

(b) Private datacenters tend to be located in places where the company or unit was founded or acquired

(c) A network backbone is a very low-capacity network connection

(d) None of the mentioned

6. Which of the following captive area deals with monitoring ?

(a) Licensed

(b) Variable but under control

(c) Low

(d) All of the above

7. Network bottlenecks occur when _____ data sets must be transferred.

(a) large

(b) small

(c) big

(d) All of the above

8. The reputation for Cloud computing services for the quality of those services is shared by _____

(a) replicas

(b) shards

(c) tenants

(d) All of the above

9. Cloud _____ are standardized in order to appeal to the majority of its audience.

(a) SVAs

(b) SLAs

(c) SALs

(d) None of the above

10. _____ is a function of the particular enterprise and application in an on-premises deployment.

(a) Vendor lock

(b) Vendor lock-in

(c) Vendor lock-ins

(d) None of the above

11. Which does C in FCAPS stands for ?

(a) Consistency

(b) Contact

(c) Configuration

(d) None of the above

12. Point out the wrong statement:

(a) Cloud computing deployments must be monitored and managed in order to be optimized for best performance

(b) To the problems associated with analyzing distributed network applications, the Cloud adds the complexity of virtual infrastructure

(c) Cloud management software provides capabilities for managing faults

(d) None of the above

13. Which of the following aims to deploy methods for measuring various aspects of Cloud performance in a standard way ?

(a) RIM

(b) SIM

(c) SMI

(d) All of the above

14. Which of the following is not the feature of Network management systems?

(a) Accounting

(b) Security

(c) Performance

(d) None of the above

15. Point out the correct statement:

(a) Cloud management includes not only managing resources in the Cloud, but managing resources on-premises

(b) The management of resources in the Cloud requires new technology

(c) Management of resources on-premises allows vendors to use well-established network management technologies

(d) All of the above

Objective type questions -answer

1:a   2:b   3:b   4:c   5:c   6:b   7:a   8:c   9:b   10:b   11:c   12:d   13:c   14:d   15:d

Review questions

1. Are there any consumer protection issues that may be relevant for Cloud computing services?

2. Which law is applicable in the case of a dispute concerning data protection and Cloud computing?

3. How the providers apply laws and regulations that are applicable to Cloud computing?

4. Who is the data controller in a Cloud computing service?

5. What are the security requirements connected with processing data?

6. Can Internet Service Providers be held liable for infringing material placed in the Cloud by their users? If yes, what are the possible sanctions for the ISP?

7. Are there any other special Intellectual Property law issues that can be relevant to Cloud computing (e.g. unusual termination of the license provisions)?

8. Several technical challenges related to the use of Cloud computing include resource exhaustion, unpredictability of performance, data lock-in, data transfer bottlenecks, and bugs in large-scale distributed Cloud systems. Comment on it.

9. Who is the owner of the data placed in the Cloud and then modified somehow by the tools available in the Cloud (e.g. formatted by the open program in the Cloud)? Who is also the owner of copies of documents placed in the Cloud (e.g. copies of sent e-mails)?

10. Are Cloud services providers free from content filtration or censorship obligations?

Critical thinking questions

1. The designers of the Google file system (GFS) have re-examined the traditional choices for a file system. Discuss observations regarding these choices that have guided the design of GFS.

2. Identify the main security threats for the SaaS Cloud delivery model on a public Cloud. Discuss the different aspects of these threats on a public Cloud vis-a-vis the threats posed to similar services provided by a traditional service-oriented architecture running on a private infrastructure.

3. Analyze Amazon privacy policies and design a service level agreement you would sign on if you were to process confidential data using AWS.

4. Analyze the implications of the lack of trusted paths in commodity operating systems and give one or more examples showing the effects of this deficiency. Analyze the implications of the two-level security model of commodity operating systems.

Bibliography

[1] X. Zhang, N. Wuwong, H. Li, and X. J. Zhang, "Information Security Risk Management Framework for the Cloud Computing Environments", *Proceedings of 10th IEEE International Conference on Computer and Information Technology*, pp. 1328-1334, 2010.

[2] Cong Wang, Qian Wang, Kui Ren, and Wenjing Lou, "Ensuring Data Storage Security in Cloud Computing", *Proceedings of 17th International workshop on Quality of Service*, USA, pp. 1-9, July 13-15, 2009.

[3] Hanqian Wu, Yi Ding, Winer, C., Li Yao, "Network Security for Virtual Machines in Cloud Computing", *Proceedings of 5th International Conference on Computer Sciences and Convergence Information Technology*, pp. 18-21, Seoul, Nov. 30-Dec. 2, 2010.

[4] S. Subashini, V. Kavitha, "A survey on security issues in service delivery models of Cloud computing", *Journal of Network and Computer Applications*, Vol. 34(1), pp 1-11, Academic Press Ltd., UK, 2011.

[5] V. Krishna Reddy, B. Thirumal Rao, Dr. L.S.S. Reddy, P.Sai Kiran, "Research Issues in Cloud Computing", *Global Journal of Computer Science and Technology*, Volume 11, Issue 11, July 2011.

A

Appendix A: Experiments Using CloudSim

Cloud computing has advanced incredibly in the ongoing years because of its characteristics of being secure, and highly scalable. We can say that it is able to dynamically deliver IT assets over the Internet. This Appendix gives a practical approach to implement and test Cloud scenarios using CloudSim simulator .

A.1 CloudSim Architecture

Figure A.1 shows the multi-layered plan of the CloudSim programming structure and its building parts. Initial arrivals of CloudSim utilized SimJava as the discrete events simulation design that bolsters a few center functionalities, for example, processing of events (administrations, have, server farm, agent, VMs), correspondence among components and the management of simulation clock. A discrete-event simulation (DES) models the activity of a framework as a (discrete) sequence of occasions in time. Every event happens at a specific moment in time and denotes a difference of state in the framework. Between continuous events, no adjustment in the framework is accepted to happen; in this manner the simulation time can legitimately hop to the event time of the next event. In the current version, the SimJava layer has been removed so as to permit some advanced operations. The CloudSim simulation layer offers help for demonstrating and simulation of virtualized Cloud-based server data center conditions including committed administration interfaces for VMs, memory, storage, and data transmission. The central issues, for example, provisioning of hosts to VMs, and managing application execution state are dealt with by this layer.

The data center deals with various host elements. The hosts are mapped out to at least one VMs dependent on a VM allotment strategy that ought to be characterized by the Cloud specialist organization. Here, the VM strategy represents the tasks control arrangements identified with VM life cycle, for example, provisioning of a host to a VM, VM creation, VM failure and VM migration. Thus, at least one application administrations can be provisioned inside a solitary VM case, alluded to as application provisioning with regards to Cloud computing.

In the context of CloudSim, an entity is an instance of a component. A CloudSim component can be a class (abstract or complete) or set of classes that represent one CloudSim model (data center, host etc.). A data center can manage several hosts that in turn manages VMs during their life cycles. Host is a CloudSim component that represents a physical computing server in a Cloud: it is assigned a pre-configured processing capability (expressed in millions of instructions per second-MIPS), memory, storage, and a provisioning policy for allocating processing cores to VMs. The Host component implements interfaces that support modeling and simulation of both single-core and multi-core nodes.

For each Host component, the allocation of processing cores to VMs is done based on a host allocation policy. This policy takes into account several hardware characteristics, such as number of CPU cores, CPU share and amount of memory (physical and secondary), that are allocated to a given VM instance. Hence, CloudSim supports simulation scenarios that assign specific CPU cores to specific VMs (a space-shared policy), dynamically distribute the capacity of a core among VMs (time-shared policy), or assign cores to VMs on demand.

CloudSim runs on any Integrated Development Environment (IDE). Let us see its intallation steps on Eclipse IDE.

Installation of CloudSim in Eclipse IDE

The installation steps are as follows:

1. Open up Eclipse and Click on java project

2. Enter project name

3. In the next line, the path for the project will be created

4. Select the JRE environment

5. Give project Name and select run time environment and Finish

6. Next Step is to go to the directory where you have extracted your Cloudsim tool.

7. Select Cloudsim and Hit finish

8. Go to the link http://commons.apache.org/proper/commonsmath/download_math.cgi.

9. Download the file named as "commons-math3-3.4.1-bin.zip". Unzip this file. We need jar files for math functions.

10. Now go to the left side of the eclipse tool in the project bar. Go to jar and right click on it. Click import, and import jar files for math calculations

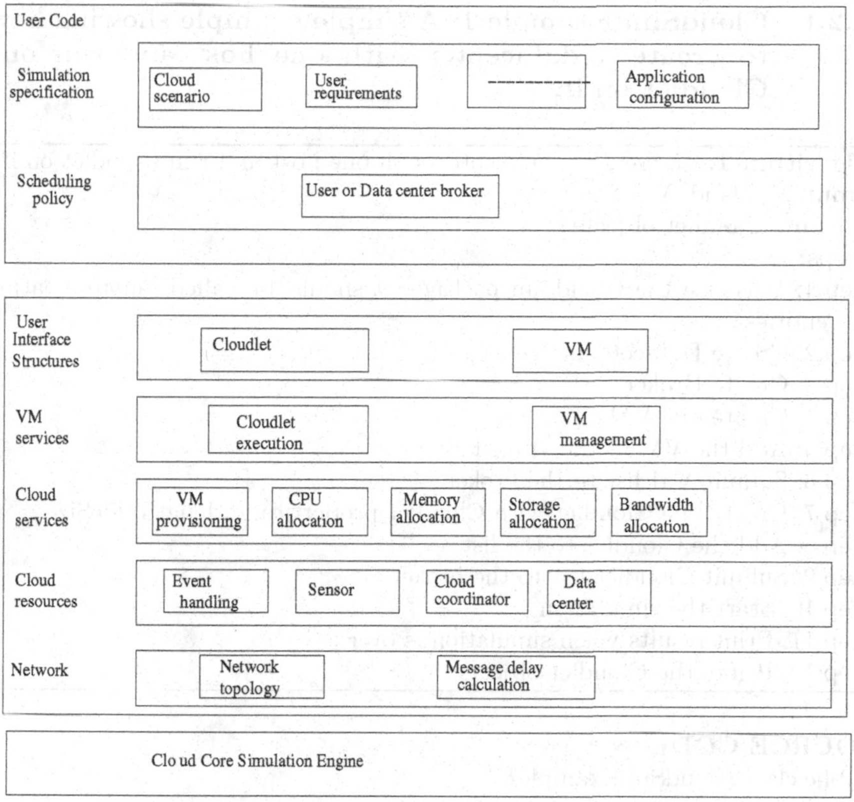

FIGURE A.1
Layered CloudSim Architecture

11. Now go to the folder where you have placed the downloaded and extracted file. Then select "Import only jar"

12. Finally the CloudSim is installed into your Eclipse environment.

In this Appendix, we consider different scenarios in Cloud and program them to achieve desired outcome using CloudSim simulator.

A.2 CloudSim Examples

Let us see some examples on CloudSim covering different scenarios of Cloud environment.

A.2.1 CloudSim Example 1: A simple example showing how to create a datacenter with one host and run one Cloudlet on it.

Algorithm 1: To create a datacenter with one host and run Cloudlet on it
Inputs: VMs id, VM Size
Output: Cloudlet objects
Steps:
Step 1: Initialize the CloudSim package. It should be called before creating any entities
Step 2: Create Datacenters
Step 3: Create Broker
Step 4: Create one VM
Step 5: Add the VM to the vmList
Step 6: Submit VM list to the broker
Step 7: Create one Cloudlet with Cloudlet properties id, length, fileSize
Step 8: Add the Cloudlet to the list
Step 9: Submit Cloudlet list to the broker
Step 10: Start the simulation
Step 11: Print results when simulation is over
Step 12: Prints the Cloudlet objects

SOURCE CODE

```
public class CloudSimExample1
{
private static List<Cloudlet> CloudletList;
private static List< Vm > vmlist;
public static void main(String[] args)
{
try {
/* First step: Initialize the CloudSim package. It should be called before creating any entities. */
int num_user = 1; // number of Cloud users
Calendar calendar = Calendar.getInstance();
boolean trace_flag = false; // trace events
CloudSim.init(num_user, calendar, trace_flag);

/* Second step: Create Datacenters */
Datacenter datacenter0 = createDatacenter("Datacenter_0");

/* Third step: Create Broker */
DatacenterBroker broker = createBroker();
int brokerId = broker.getId();
```

```
/* Fourth step: Create one virtual machine */
vmlist = new ArrayList< Vm >();

/* VM description * /
int vmid = 0;
int mips = 1000;
long size = 10000; // image size (MB)
int ram = 512; // vm memory (MB)
long bw = 1000;
int pesNumber = 1; // number of cpus
String vmm = "Xen"; // VMM name

/* create VM */
Vm vm = new Vm(vmid, brokerId, mips, pesNumber, ram, bw, size, vmm,
new CloudletSchedulerTimeShared());

/* add the VM to the vmList */
vmlist.add(vm);

/* submit vm list to the broker */
broker.submitVmList(vmlist);

/* Fifth step: Create one Cloudlet */
CloudletList = new ArrayList< Cloudlet >();

/* Cloudlet properties */
int id = 0;
length = 400000;
fileSize = 300;
outputSize = 300;
utilizationModel = new UtilizationModelFull();
Cloudlet Cloudlet = new Cloudlet(id, length, pesNumber, fileSize,outputSize,
utilizationModel, utilizationModel,utilizationModel);
Cloudlet.setUserId(brokerId);
Cloudlet.setVmId(vmid);

/* add the Cloudlet to the list */
CloudletList.add(Cloudlet);

/* submit Cloudlet list to the broker */
broker.submitCloudletList(CloudletList);

/* Sixth step: Starts the simulation */
CloudSim.startSimulation();
CloudSim.stopSimulation();
```

```
/* Final step: Print results when simulation is over */ List< Cloudlet >
newList = broker.getCloudletReceivedList();
printCloudletList(newList);
Log.printLine("CloudSimExample1 finished!");
}

catch (Exception e)
{
e.printStackTrace();
Log.printLine("Unwanted errors happen");
}
}

/* Create the datacenter */

private static Datacenter createDatacenter(String name) {

/* Here are the steps needed to create a PowerDatacenter: */
/* 1. We need to create a list to store our machine */
List< Host > hostList = new ArrayList< Host >();

/* 2. A Machine contains one or more PEs or CPUs/Cores. In this exam-
ple, it will have only one core. */
List< Pe > peList = new ArrayList< Pe >();
int mips = 1000;

/* 3. Create PEs and add these into a list. */
peList.add(new Pe(0, new PeProvisionerSimple(mips))); // need to store Pe
id and MIPS Rating

/* 4. Create Host with its id and list of PEs and add them to the list */
/* of machines */

int hostId = 0;
int ram = 2048; // host memory (MB)
long storage = 1000000; // host storage
int bw = 10000;

hostList.add(new Host(hostId,new RamProvisionerSimple(ram),new BwPro-
visionerSimple(bw),storage,peList,new VmSchedulerTimeShared(peList)));
/* This is our machine's configuration */
/* 5. Create a Datacenter Characteristics object that stores the properties of
a data center: architecture, OS, list of Machines, allocation
policy: time or space-shared, time zone and its price (GPe time unit).
```

String arch = "x86"; // system architecture
String os = "Linux"; // operating system
String vmm = "Xen";
double time_zone = 10.0; // time zone this resource located
double cost = 3.0; // the cost of using processing in this resource
double costPerMem = 0.05; // the cost of using memory in this resource
double costPerStorage = 0.001; // the cost of using storage in this
// resource
double costPerBw = 0.0; // the cost of using bw in this resource

LinkedList< *Storage* > storageList = new LinkedList< *Storage* >(); // we are not adding SAN

DatacenterCharacteristics characteristics = new DatacenterCharacteristics(arch, os, vmm, hostList, time_zone, cost, costPerMem, costPerStorage, costPerBw);

/* 6. Finally, we need to create a PowerDatacenter object. */

Datacenter datacenter = null;

try
{
datacenter = new Datacenter(name, characteristics, new VmAllocationPolicySimple(hostList), storageList, 0);
}

    catch (Exception e)
{
e.printStackTrace();
}
return datacenter;
/* Create the broker */
private static DatacenterBroker createBroker()
{
DatacenterBroker broker = null;
try {
broker = new DatacenterBroker("Broker");
}
catch (Exception e) {
e.printStackTrace();
return null;
}
return broker; }

/* Prints the Cloudlet objects */

```
private static void printCloudletList(List< Cloudlet > list)
{
int size = list.size();
Cloudlet Cloudlet;
Log.printLine("=========== OUTPUT ==========");
Log.printLine("Cloudlet ID" + "STATUS" + "Data center ID" + "VM ID"
+ "Time" + "Start Time" + "Finish Time");
for (int i = 0; i < size; i++)
{
Cloudlet = list.get(i);
Log.print(Cloudlet.getCloudletId());
if (Cloudlet.getCloudletStatus() == Cloudlet.SUCCESS)
{
Log.print("SUCCESS");
Log.printLine(Cloudlet.getResourceId() + Cloudlet.getVmId() + Cloudlet.
getActualCPUTime()) + Cloudlet.getExecStartTime())
+ Cloudlet.getFinishTime());
}
}
}
}
```

Output: CloudSimExample1 Starting CloudSimExample1...

Initialising...

Starting CloudSim version 3.0

Datacenter_0 is starting...

Broker is starting...

Entities started.

0.0: Broker: Cloud Resource List received with 1 resource(s)

0.0: Broker: Trying to Create VM #0 in Datacenter_0

0.1: Broker: VM #0 has been created in Datacenter #2, Host #0

0.1: Broker: Sending Cloudlet 0 to VM #0

400.1: Broker: Cloudlet 0 received

400.1: Broker: All Cloudlets executed. Finishing...

400.1: Broker: Destroying VM #0

Broker is shutting down..

. Simulation: No more future events

CloudInformationService: Notify all CloudSim entities for shutting down.

Datacenter_0 is shutting down...

Broker is shutting down...

Simulation completed.

TABLE A.1
Output: Example 1

Cloudlet ID	STATUS	Data Center ID	VM ID	Time (ms)	Start Time	Finish Time
0	success	2	0	400	0.1	400.1

A.2.2 CloudSim Example 2: A simple example showing how to create two datacenters with one host and a network topology each and run two Cloudlets on them.

Algorithm 2: To create two datacenters with one host and a network topology each and run two Cloudlets on them

Inputs: Two VMs id, VM Size, PEs configurations

Output: Two Cloudlet objects

Steps:

Step 1: Initialize the CloudSim package. It should be called before creating any entities

Step 2: Initialize the CloudSim library

Step 3: Create Datacenters

Step 4: Create Broker

Step 5: Create two VMs with VM_id, size, ram

Step 6: Add the VMs to the vmList

Step 7: Submit VM list to the broker

Step 8: Create two Cloudlets: Cloudlet properties, length, fileSize

Step 9: Add the Cloudlets to the list

Step 10: Submit Cloudlet list to the broker

Step 11: Bind the Cloudlets to the vms. This way, the broker will submit the bound Cloudlets only to the specific VM

Step 12: Starts the simulation

Step 13: Print results when simulation is over

Step 14: Create Host with its id and list of PEs and add them to the list of machines

Step 15: Create a DatacenterCharacteristics object that stores the properties of a data center: architecture, OS, list of

Machines, allocation policy: time or space-shared, time zone and its price (G$/Pe time unit).

Step 16: Create a PowerDatacenter object.

Step 17: Prints the Cloudlet objects

SOURCE CODE

public class CloudSimExample2

```
/** The Cloudlet list. */
private static List< Cloudlet > CloudletList;

/** The vmlist. */
private static List< Vm > vmlist;

public static void main(String[] args)

try

// First step: Initialize the CloudSim package. It should be called
// before creating any entities.

int num_user = 1; // number of Cloud users
Calendar calendar = Calendar.getInstance();
boolean trace_flag = false; // mean trace events

// Initialize the CloudSim library CloudSim.init(num_user, calendar,
trace_flag);

// Second step: Create Datacenters
Datacenter datacenter0 = createDatacenter("Datacenter_0");

//Third step: Create Broker
DatacenterBroker broker = createBroker();
int brokerId = broker.getId();

//Fourth step: Create one virtual machine vmlist = new ArrayList< Vm >();

//VM description
int vmid = 0;
int mips = 250;
long size = 10000; //image size (MB)
int ram = 512; //vm memory (MB)
long bw = 1000;
int pesNumber = 1; //number of cpus
String vmm = "Xen"; //VMM name

//create two VMs
Vm vm1 = new Vm(vmid, brokerId, mips, pesNumber, ram, bw, size, vmm,
new CloudletSchedulerTimeShared());
```

```
vmid++;
Vm vm2 = new Vm(vmid, brokerId, mips, pesNumber, ram, bw, size, vmm,
new CloudletSchedulerTimeShared());

//add the VMs to the vmList
vmlist.add(vm1);
vmlist.add(vm2);

//submit vm list to the broker
broker.submitVmList(vmlist);

//Fifth step: Create two Cloudlets
CloudletList = new ArrayList< Cloudlet >();

//Cloudlet properties
int id = 0;
pesNumber=1;
long length = 250000;
long fileSize = 300;
long outputSize = 300;
UtilizationModel utilizationModel = new UtilizationModelFull();
Cloudlet Cloudlet1 = new Cloudlet(id, length, pesNumber, fileSize, output-
Size, utilizationModel, utilizationModel, utilizationModel);
Cloudlet1.setUserId(brokerId);
id++;
Cloudlet Cloudlet2 = new Cloudlet(id, length, pesNumber, fileSize, output-
Size, utilizationModel, utilizationModel, utilizationModel);
Cloudlet2.setUserId(brokerId);

//add the Cloudlets to the list
CloudletList.add(Cloudlet1);
CloudletList.add(Cloudlet2);

//submit Cloudlet list to the broker
broker.submitCloudletList(CloudletList);

//bind the Cloudlets to the vms. This way, the broker
// will submit the bound Cloudlets only to the specific VM
broker.bindCloudletToVm(Cloudlet1.getCloudletId(),vm1.getId());
broker.bindCloudletToVm(Cloudlet2.getCloudletId(),vm2.getId());

// Sixth step: Starts the simulation
CloudSim.startSimulation();

// Final step: Print results when simulation is over
```

```
List< Cloudlet > newList = broker.getCloudletReceivedList();
CloudSim.stopSimulation();
printCloudletList(newList);
Log.printLine("CloudSimExample2 finished!");

catch (Exception e)
e.printStackTrace();
Log.printLine("The simulation has been terminated due to an unexpected er-
ror");


    private static Datacenter createDatacenter(String name)

// Here are the steps needed to create a PowerDatacenter:
// 1. We need to create a list to store our machine
List< Host > hostList = new ArrayList< Host >();

// 2. A Machine contains one or more PEs or CPUs/Cores.
// In this example, it will have only one core.
List< Pe > peList = new ArrayList< Pe >();
int mips = 1000;

// 3. Create PEs and add these into a list.
peList.add(new Pe(0, new PeProvisionerSimple(mips))); // need to store Pe
id and MIPS Rating

//4. Create Host with its id and list of PEs and add them to the list of
machines
int hostId=0;
int ram = 2048; //host memory (MB)
long storage = 1000000; //host storage (MB)
int bw = 10000; //bandwidth (Mbps)

hostList.add(
new Host(
hostId,
new RamProvisionerSimple(ram),
new BwProvisionerSimple(bw),
storage,
peList,
new VmSchedulerTimeShared(peList))
); // This is user's machine
```

```
// 5. Create a DatacenterCharacteristics object that stores the
// properties of a data center: architecture, OS, list of
// Machines, allocation policy: time or space-shared, time zone
// and its price (G$/Pe time unit).
String arch = "x86"; // system architecture
String os = "Linux"; // operating system
String vmm = "Xen";
// virtual machine monitor
double time_zone = 10.0; // time zone this resource located
double cost = 3.0; // the cost of using processing in this resource
double costPerMem = 0.05; // the cost of using memory in this resource
double costPerStorage = 0.001; // the cost of using storage in this resource
double costPerBw = 0.0; // the cost of using bw in this resource
LinkedList< Storage > storageList = new LinkedList< Storage >();
DatacenterCharacteristics characteristics = new DatacenterCharacteristics(
arch, os, vmm, hostList, time_zone, cost, costPerMem, costPerStorage, cost-
PerBw);


// 6. Finally, we need to create a PowerDatacenter object. Datacenter data-
center = null;
try
datacenter = new Datacenter(name, characteristics, new VmAllocationPoli-
cySimple(hostList), storageList, 0);
catch (Exception e)
e.printStackTrace();

return datacenter;


//We strongly encourage users to develop their own broker policies, to submit
vms and Cloudlets according
//to the specific rules of the simulated scenario
private static DatacenterBroker createBroker()
DatacenterBroker broker = null;
try
broker = new DatacenterBroker("Broker");
catch (Exception e)
e.printStackTrace();
return null;

return broker;


/** Prints the Cloudlet objects*//
```

```
private static void printCloudletList(List< Cloudlet > list)
int size = list.size();
Cloudlet Cloudlet;
Log.printLine("========== OUTPUT ==========");
Log.printLine("Cloudlet ID" + "STATUS" + "Data center ID" + "VM ID"
+ "Time" + "Start Time" + "Finish Time");
for (int i = 0; i < size; i++)
Cloudlet = list.get(i);
Log.print(+ Cloudlet.getCloudletId());

if (Cloudlet.getCloudletStatus() == Cloudlet.SUCCESS)
Log.print("SUCCESS");

Log.printLine(Cloudlet.getResourceId() + Cloudlet.getVmId() + (Cloudlet.
getActualCPUTime()) +
Cloudlet.getExecStartTime())+ + (Cloudlet.getFinishTime());
```

Output: CloudSimExample2 Starting CloudSimExample2...
Initialising...
Starting CloudSim version 3.0
Datacenter_0 is starting...
Broker is starting...
Entities started.
0.0: Broker: Cloud Resource List received with 1 resource(s)
0.0: Broker: Trying to Create VM #0 in Datacenter_0
0.0: Broker: Trying to Create VM #1 in Datacenter_0
0.1: Broker: VM #0 has been created in Datacenter #2, Host #0
0.1: Broker: VM #1 has been created in Datacenter #2, Host #0
0.1: Broker: Sending Cloudlet 0 to VM #0
0.1: Broker: Sending Cloudlet 1 to VM #1
1000.1: Broker: Cloudlet 0 received
1000.1: Broker: Cloudlet 1 received
1000.1: Broker: All Cloudlets executed. Finishing...
1000.1: Broker: Destroying VM #0
1000.1: Broker: Destroying VM #1
Broker is shutting down...
Simulation: No more future events
CloudInformationService: Notify all CloudSim entities for shutting down.
Datacenter_0 is shutting down...
Broker is shutting down...
Simulation completed.

TABLE A.2
Output: Example 2

Cloudlet ID	STATUS	Data Center ID	VM ID	Time	Start Time	Finish Time
0	success	2	0	1000	0.1	1000.1
1	success	2	0	1000	0.1	1000.1

A.2.3 CloudSim Example 3: A simple example showing how to create two datacenters with one host each and run Cloudlets of two users with network topology on them.

Algorithm 3: To create two datacenters with one host each and run Cloudlets of two users with network topology on them
Inputs: Two VMs id, VM Size, PEs configurations
Output: User Cloudlets
Steps:
Step 1: Initialize the CloudSim package. It should be called before creating any entities.
Step 2: Create Datacenters: Datacenters are the resource providers in CloudSim. We need at list one of them to run a CloudSim simulation
Step 3: Create Broker
Step 4: Create one VM machine
Step 5: Create two VMs
Step 6: The second VM will have twice the priority of VM1 and so will receive twice CPU time
Step 7: Add the VMs to the vmList
Step 8: Submit VM list to the broker
Step 9: Create two Cloudlets
Step 10: Submit Cloudlet list to the broker
Step 11: Bind the Cloudlets to the vms. This way, the broker will submit the bound Cloudlets only to the specific VM
Step 12: Starts the simulation
Step 13: Prints result when simulation is over
Step 14: Create a PowerDatacenter
Step 15: Create PEs and add these into a list
Step 16: Create Hosts with its id and list of PEs and add them to the list of machines
Step 17: Create a DatacenterCharacteristics object that stores the properties of a data center: architecture, OS, list of properties
Step 18: Finally, we need to create a PowerDatacenter object

SOURCE CODE

public class CloudSimExample3

```
/** The Cloudlet list. */
private static List< Cloudlet > CloudletList;

/** The vmlist. */
private static List< Vm > vmlist;

public static void main(String[] args)
Log.printLine("Starting CloudSimExample3...");
try
// First step: Initialize the CloudSim package. It should be called
// before creating any entities.
int num_user = 1; // number of Cloud users
Calendar calendar = Calendar.getInstance();
boolean trace_flag = false; // mean trace events
// Initialize the CloudSim library
CloudSim.init(num_user, calendar, trace_flag);
// Second step: Create Datacenters
//Datacenters are the resource providers in CloudSim. We need at list one of
them to run a CloudSim simulation
Datacenter datacenter0 = createDatacenter("Datacenter_0");
//Third step: Create Broker
DatacenterBroker broker = createBroker();
int brokerId = broker.getId();
//Fourth step: Create one virtual machine
vmlist = new ArrayList< Vm >();

//VM description
int vmid = 0;
int mips = 250;
long size = 10000; //image size (MB)
int ram = 2048; //vm memory (MB)
long bw = 1000;
int pesNumber = 1; //number of cpus
String vmm = "Xen"; //VMM name

//create two VMs
Vm vm1 = new Vm(vmid, brokerId, mips, pesNumber, ram, bw, size, vmm,
new CloudletSchedulerTimeShared());

//the second VM will have twice the priority of VM1 and so will receive
twice CPU time
```

```
vmid++;
Vm vm2 = new Vm(vmid, brokerId, mips * 2, pesNumber, ram, bw, size,
vmm, new CloudletSchedulerTimeShared());

//add the VMs to the vmList
vmlist.add(vm1);
vmlist.add(vm2);

//submit vm list to the broker
broker.submitVmList(vmlist);

//Fifth step: Create two Cloudlets
CloudletList = new ArrayList< Cloudlet >();

//Cloudlet properties
int id = 0;
long length = 40000;
long fileSize = 300;
long outputSize = 300;
UtilizationModel utilizationModel = new UtilizationModelFull();

Cloudlet Cloudlet1 = new Cloudlet(id, length, pesNumber, fileSize, output-
Size, utilizationModel, utilizationModel, utilizationModel);
Cloudlet1.setUserId(brokerId);
id++;
Cloudlet Cloudlet2 = new Cloudlet(id, length, pesNumber, fileSize, output-
Size, utilizationModel, utilizationModel, utilizationModel);
Cloudlet2.setUserId(brokerId);

//add the Cloudlets to the list
CloudletList.add(Cloudlet1);
CloudletList.add(Cloudlet2);

//submit Cloudlet list to the broker
broker.submitCloudletList(CloudletList);

//bind the Cloudlets to the vms. This way, the broker
// will submit the bound Cloudlets only to the specific VM
broker.bindCloudletToVm(Cloudlet1.getCloudletId(),vm1.getId());
broker.bindCloudletToVm(Cloudlet2.getCloudletId(),vm2.getId());

// Sixth step: Starts the simulation
CloudSim.startSimulation();

// Final step: Prints result when simulation is over
```

```
List< Cloudlet > newList = broker.getCloudletReceivedList();
CloudSim.stopSimulation();
printCloudletList(newList);
Log.printLine("CloudSimExample3 finished!");

catch (Exception e)
e.printStackTrace();
Log.printLine("The simulation has been terminated due to an unexpected er-
ror");


private static Datacenter createDatacenter(String name)

// Here are the steps needed to create a PowerDatacenter:
// 1. We need to create a list to store
// our machine
List< Host > hostList = new ArrayList< Host >();

// 2. A Machine contains one or more PEs or CPUs/Cores.
// In this example, it will have only one core.
List< Pe > peList = new ArrayList< Pe >();
int mips = 1000;

// 3. Create PEs and add these into a list.
peList.add(new Pe(0, new PeProvisionerSimple(mips))); // need to store Pe
id and MIPS Rating

//4. Create Hosts with its id and list of PEs and add them to the list of
machines
int hostId=0;
int ram = 2048; //host memory (MB)
long storage = 1000000; //host storage
int bw = 10000;
hostList.add(
new Host(
hostId,
new RamProvisionerSimple(ram),
new BwProvisionerSimple(bw),
storage,
peList,
new VmSchedulerTimeShared(peList)
)
); // This is our first machine

//create another machine in the Data center
```

```
List< Pe > peList2 = new ArrayList< Pe >();

peList2.add(new Pe(0, new PeProvisionerSimple(mips)));
hostId++;
hostList.add(
new Host(
hostId,
new RamProvisionerSimple(ram),
new BwProvisionerSimple(bw),
storage,
peList2,
new VmSchedulerTimeShared(peList2)
)
); // This is our second machine

// 5. Create a DatacenterCharacteristics object that stores the properties
of a data center: architecture, OS, list of
// Machines, allocation policy: time- or space-shared, time zone
// and its price (G$/Pe time unit).
String arch = "x86"; // system architecture
String os = "Linux"; // operating system
String vmm = "Xen";
double time_zone = 10.0; // time zone this resource located
double cost = 3.0; // the cost of using processing in this resource
double costPerMem = 0.05; // the cost of using memory in this resource
double costPerStorage = 0.001; // the cost of using storage in this resource
double costPerBw = 0.0; // the cost of using bw in this resource
LinkedList< Storage > storageList = new LinkedList< Storage >(); //we
are not adding SAN devices by now

    DatacenterCharacteristics characteristics = new DatacenterCharacteristics(
arch, os, vmm, hostList, time_zone, cost, costPerMem, costPerStorage, costPerBw);

// 6. Finally, we need to create a PowerDatacenter object.
Datacenter datacenter = null;
try
datacenter = new Datacenter(name, characteristics, new VmAllocationPolicySimple(hostList), storageList, 0);
catch (Exception e)
e.printStackTrace();
```

```
        return datacenter;


    private static DatacenterBroker createBroker()

    DatacenterBroker broker = null;
try
broker = new DatacenterBroker("Broker");
catch (Exception e)
e.printStackTrace();
return null;
return broker;



/**
 * Prints the Cloudlet objects */
private static void printCloudletList(List< Cloudlet > list)
int size = list.size();
Cloudlet Cloudlet;
Log.printLine("=========== OUTPUT ===========");
Log.printLine("Cloudlet ID" + "STATUS" + "Data center ID" + "VM ID"
+ "Time" +"Start Time" + "Finish Time");
for (int i = 0; i < size; i++)
Cloudlet = list.get(i);
Log.print(Cloudlet.getCloudletId());

if (Cloudlet.getCloudletStatus() == Cloudlet.SUCCESS)
Log.print("SUCCESS");
Log.printLine(Cloudlet.getResourceId() + Cloudlet.getVmId() + Cloudlet.
getActualCPUTime() +
Cloudlet.getExecStartTime()+ Cloudlet.getFinishTime());
```

Output: CloudSimExample 3

Starting CloudSimExample3...
Initialising...
Starting CloudSim version 3.0
Datacenter_0 is starting...
Broker is starting...
Entities started.
0.0: Broker: Cloud Resource List received with 1 resource(s)
0.0: Broker: Trying to Create VM #0 in Datacenter_0
0.0: Broker: Trying to Create VM #1 in Datacenter_0
0.1: Broker: VM #0 has been created in Datacenter #2, Host #0
0.1: Broker: VM #1 has been created in Datacenter #2, Host #1
0.1: Broker: Sending Cloudlet 0 to VM #0
0.1: Broker: Sending Cloudlet 1 to VM #1
80.1: Broker: Cloudlet 1 received
160.1: Broker: Cloudlet 0 received
160.1: Broker: All Cloudlets executed. Finishing...
160.1: Broker: Destroying VM #0
160.1: Broker: Destroying VM #1
Broker is shutting down...
Simulation: No more future events
CloudInformationService: Notify all CloudSim entities for shutting down.
Datacenter_0 is shutting down...
Broker is shutting down...
Simulation completed.

TABLE A.3
Output: Example 3

Cloudlet ID	STATUS	Data Center ID	VM ID	Time	Start Time	Finish Time
1	SUCCESS	2	1	80	0.1	80.1
0	SUCCESS	2	0	160	0.1	160.1

A.2.4   CloudSim Example 4: A simple example showing how to create two datacenters with one host each and run two Cloudlets on them

Algorithm 4: To create two datacenters with one host each and run two Cloudlets on them

Inputs: Two VMs id, VM Size, PEs configurations

Output: User Cloudlets

Steps:

Step 1: Initialize the CloudSim package. It should be called before creating any entities

Step 2: Create Datacenters

Step 3: Create Broker

Step 4: Create one virtual machine

Step 5: Create two VMs

Step 6: Add the VMs to the vmList

Step 7: Submit vm list to the broker

Step 8: Create two Cloudlets

Step 9: Add the Cloudlets to the list

Step 10: Submit Cloudlet list to the broker

Step 11: Bind the Cloudlets to the vms. This way, the broker will submit the bound Cloudlets only to the specific VM

Step 12: Starts the simulation

Step 13: Print results when simulation is over

Step 14: Create a PowerDatacenter

Step 15: Create PEs and add these into a list

Step 16: Create Host with its id and list of PEs and add them to the list of machines

Step 17: Create a DatacenterCharacteristics object that stores the properties of a data center: architecture, OS, list of Machines, allocation policy: time- or space-shared, time zone and its price (G\$/Pe time unit).

Step 18: Finally, we need to create a PowerDatacenter object

Step 19: Prints the User Cloudlet

SOURCE CODE

public class CloudSimExample4

```
/** The Cloudlet list. */ private static List< Cloudlet > CloudletList;
/** The vmlist. */
private static List< Vm > vmlist;
public static void main(String[] args)
Log.printLine("Starting CloudSimExample4...");
try
// First step: Initialize the CloudSim package. It should be called
// before creating any entities.
int num_user = 1; // number of Cloud users
```

```
Calendar calendar = Calendar.getInstance();
boolean trace_flag = false; // mean trace events

CloudSim.init(num_user, calendar, trace_flag);

// Second step: Create Datacenters
//Datacenters are the resource providers in CloudSim. We need at list one of
them to run a CloudSim simulation
Datacenter datacenter0 = createDatacenter("Datacenter_0");
Datacenter datacenter1 = createDatacenter("Datacenter_1");

//Third step: Create Broker
DatacenterBroker broker = createBroker();
int brokerId = broker.getId();

//Fourth step: Create one virtual machine
vmlist = new ArrayList< Vm >();

//VM description
int vmid = 0;
int mips = 250;
long size = 10000; //image size (MB)
int ram = 512; //vm memory (MB)
long bw = 1000;
int pesNumber = 1; //number of cpus
String vmm = "Xen"; //VMM name

//create two VMs
Vm vm1 = new Vm(vmid, brokerId, mips, pesNumber, ram, bw, size, vmm,
new CloudletSchedulerTimeShared());

vmid++;
Vm vm2 = new Vm(vmid, brokerId, mips, pesNumber, ram, bw, size, vmm,
new CloudletSchedulerTimeShared());

//add the VMs to the vmList
vmlist.add(vm1);
vmlist.add(vm2);

//submit vm list to the broker
broker.submitVmList(vmlist);

//Fifth step: Create two Cloudlets
CloudletList = new ArrayList< Cloudlet >();
```

```
//Cloudlet properties
int id = 0;
long length = 40000;
long fileSize = 300;
long outputSize = 300;
UtilizationModel utilizationModel = new UtilizationModelFull();

Cloudlet Cloudlet1 = new Cloudlet(id, length, pesNumber, fileSize, output-
Size, utilizationModel, utilizationModel,
utilizationModel);
Cloudlet1.setUserId(brokerId);

id++;
Cloudlet Cloudlet2 = new Cloudlet(id, length, pesNumber, fileSize, output-
Size, utilizationModel, utilizationModel,
utilizationModel);
Cloudlet2.setUserId(brokerId);

//add the Cloudlets to the list
CloudletList.add(Cloudlet1);
CloudletList.add(Cloudlet2);

//submit Cloudlet list to the broker
broker.submitCloudletList(CloudletList);

//bind the Cloudlets to the vms. This way, the broker
// will submit the bound Cloudlets only to the specific VM
broker.bindCloudletToVm(Cloudlet1.getCloudletId(),vm1.getId());
broker.bindCloudletToVm(Cloudlet2.getCloudletId(),vm2.getId());

// Sixth step: Starts the simulation
CloudSim.startSimulation();

// Final step: Print results when simulation is over
List< Cloudlet > newList = broker.getCloudletReceivedList();
CloudSim.stopSimulation();
printCloudletList(newList);
Log.printLine("CloudSimExample4 finished!");

catch (Exception e)
e.printStackTrace();
Log.printLine("The simulation has been terminated due to an unexpected er-
ror");

private static Datacenter createDatacenter(String name)
```

```
// Here are the steps needed to create a PowerDatacenter:
// 1. We need to create a list to store
// our machine
List< Host > hostList = new ArrayList< Host >();

// 2. A Machine contains one or more PEs or CPUs/Cores.
// In this example, it will have only one core.
List< Pe > peList = new ArrayList< Pe >();
int mips = 1000;

// 3. Create PEs and add these into a list.
peList.add(new Pe(0, new PeProvisionerSimple(mips))); // need to store Pe
id and MIPS Rating

//4. Create Host with its id and list of PEs and add them to the list of
machines
int hostId=0;
int ram = 2048; //host memory (MB)
long storage = 1000000; //host storage
int bw = 10000;

//in this example, the VMAllocatonPolicy in use is SpaceShared. It means
that only one VM
//is allowed to run on each Pe. As each Host has only one Pe, only one VM
can run on each Host.
hostList.add(
new Host(
hostId,
new RamProvisionerSimple(ram),
new BwProvisionerSimple(bw),
storage,
peList,
new VmSchedulerSpaceShared(peList)
)
); // This is our first machine

// 5. Create a DatacenterCharacteristics object that stores the
// properties of a data center: architecture, OS, list of
// Machines, allocation policy: time- or space-shared, time zone
// and its price (G$/Pe time unit).
String arch = "x86"; // system architecture
String os = "Linux"; // operating system
String vmm = "Xen";
double time_zone = 10.0; // time zone this resource located
```

```
double cost = 3.0; // the cost of using processing in this resource
double costPerMem = 0.05; // the cost of using memory in this resource
double costPerStorage = 0.001; // the cost of using storage in this resource
double costPerBw = 0.0; // the cost of using bw in this resource
LinkedList< Storage > storageList = new LinkedList< Storage >(); //we
are not adding SAN devices by now

    DatacenterCharacteristics characteristics = new DatacenterCharacteris-
tics(
arch, os, vmm, hostList, time_zone, cost, costPerMem, costPerStorage, cost-
PerBw);

// 6. Finally, we need to create a PowerDatacenter object.
Datacenter datacenter = null;
try
datacenter = new Datacenter(name, characteristics, new VmAllocationPoli-
cySimple(hostList), storageList, 0);
catch (Exception e)
e.printStackTrace();

return datacenter;

//We strongly encourage users to develop their own broker policies, to submit
vms and Cloudlets according
//to the specific rules of the simulated scenario

private static DatacenterBroker createBroker()
DatacenterBroker broker = null;
try
broker = new DatacenterBroker("Broker");
catch (Exception e)
e.printStackTrace();
return null;

return broker;


/** Prints the Cloudlet objects */
private static void printCloudletList(List< Cloudlet > list)
int size = list.size();
Cloudlet Cloudlet;
Log.printLine();
Log.printLine("=========== OUTPUT ===========");
Log.printLine("Cloudlet ID" + "STATUS" + "Data center ID" + "VM ID"
+ "Time" + "Start Time" + "Finish Time");
```

```
for (int i = 0; i < size; i++)
Cloudlet = list.get(i);
Log.print(Cloudlet.getCloudletId());
if (Cloudlet.getCloudletStatus() == Cloudlet.SUCCESS)
Log.print("SUCCESS");
Log.printLine( Cloudlet.getResourceId() + Cloudlet.getVmId() + Cloudlet.
getActualCPUTime() + Cloudlet.getExecStartTime()+ Cloudlet.
getFinishTime());
```

Output: CloudSim Example4
Starting CloudSimExample4...
Initialising...
Starting CloudSim version 3.0
Datacenter_0 is starting...
Datacenter_1 is starting...
Broker is starting...
Entities started.
0.0: Broker: Cloud Resource List received with 2 resource(s)
0.0: Broker: Trying to Create VM #0 in Datacenter_0
0.0: Broker: Trying to Create VM #1 in Datacenter_0
[VmScheduler.vmCreate] Allocation of VM #1 to Host #0 failed by MIPS
0.1: Broker: VM #0 has been created in Datacenter #2, Host #0
0.1: Broker: Creation of VM #1 failed in Datacenter #2
0.1: Broker: Trying to Create VM #1 in Datacenter_1
0.2: Broker: VM #1 has been created in Datacenter #3, Host #0
0.2: Broker: Sending Cloudlet 0 to VM #0
0.2: Broker: Sending Cloudlet 1 to VM #1
160.2: Broker: Cloudlet 0 received
160.2: Broker: Cloudlet 1 received
160.2: Broker: All Cloudlets executed. Finishing...
160.2: Broker: Destroying VM #0
160.2: Broker: Destroying VM #1
Broker is shutting down...
Simulation: No more future events
CloudInformationService: Notify all CloudSim entities for shutting down.
Datacenter_0 is shutting down...
Datacenter_1 is shutting down...
Broker is shutting down...
Simulation completed.

TABLE A.4

Output: Example4

Cloudlet ID	STATUS	Data Center ID	VM ID	Time	Start Time	Finish Time
0	SUCCESS	2	0	160	0.2	160.2
1	SUCCESS	3	1	160	0.2	160.2

Algorithm 5: An example exploring the problem of initial placement of containers.

Inputs: Two VMs id, VM Size, PEs configurations

Output: User Cloudlets

Steps:

Step 1: Get the path of the planet lab workload that is included in the cloudSim Package

Step 2: Create output folder for the logs. The log files would be located in this folder.

Step 3: Define the allocation policy for VMs.

Step 4: Define the selection policy for containers where a container migration is triggered.

Step 5: Define the allocation policy used for allocating containers to VMs.

Step 6: Define the host selection policy that determines which hosts should be selected as the migration destination.

Step 7: Define the VM Selection Policy that is used for selecting VMs to migrate when a host status is determined as Overloaded

Step 8: The container overbooking factor is used for overbooking resources of the VM.

A.2.5 CloudSim Example 5: An example exploring the problem of initial placement of containers.

SOURCE CODE

package org.cloudbus.cloudsim.examples.container;

public class ContainerInitialPlacementTest

public static void main(String[] args) throws IOException
/** * The experiments can be repeated for (repeat runtime +1) times.
* Please set these values as the arguments of the main function or set them
bellow:
*/

int runTime = Integer.parseInt(args[0]);
int repeat = Integer.parseInt(args[1]);
for (int i = runTime; i <repeat>; ++i)
boolean enableOutput = true;
boolean outputToFile = true;

/**
* Getting the path of the planet lab workload that is included in the cloudSim
Package
*/
String inputFolder = ContainerOverbooking.class.getClassLoader(). getRe-
source("workload/planetlab").getPath();
/**

* The output folder for the logs. The log files would be located in this folder.
*/
String outputFolder = " /Results";

/**
* The allocation policy for VMs.
*/
String vmAllocationPolicy = "MSThreshold-Under_0.80_0.70";

/**
* The selection policy for containers where a container migration is triggered.
*/
String containerSelectionPolicy = "MaxUsage";

/**

* The allocation policy used for allocating containers to VMs.
*/
String containerAllocationPolicy = "MostFull";
/**

* The host selection policy determines which hosts should be selected as the
migration destination.
*/

```
String hostSelectionPolicy = "FirstFit";

/**
* The VM Selection Policy is used for selecting VMs to migrate when a host
status is determined as
* "Overloaded"
*/
String vmSelectionPolicy = "VmMaxC";

/**
* The container overbooking factor is used for overbooking resources of the
VM. In this specific case
* the overbooking is performed on CPU only.
*/

int OverBookingFactor = 80;

new RunnerInitiator(
enableOutput,
outputToFile,
inputFolder,
outputFolder,
vmAllocationPolicy,
containerAllocationPolicy,
vmSelectionPolicy,
containerSelectionPolicy,
hostSelectionPolicy,
OverBookingFactor, Integer.toString(i), outputFolder);
```

OUTPUT: CloudSim Example5

Starting CloudSimExample5...

Initialising...

Starting CloudSim version 3.0

Datacenter_0 is starting...

Datacenter_1 is starting...

Broker1 is starting...

Broker2 is starting...

Entities started.

0.0: Broker1: Cloud Resource List received with 2 resource(s)

0.0: Broker2: Cloud Resource List received with 2 resource(s)

0.0: Broker1: Trying to Create VM #0 in Datacenter_0

0.0: Broker2: Trying to Create VM #0 in Datacenter_0

[VmScheduler.vmCreate] Allocation of VM #0 to Host #0 failed by MIPS

0.1: Broker1: VM #0 has been created in Datacenter #2, Host #0

0.1: Broker1: Sending Cloudlet 0 to VM #0

0.1: Broker2: Creation of VM #0 failed in Datacenter #2

0.1: Broker2: Trying to Create VM #0 in Datacenter_1

0.2: Broker2: VM #0 has been created in Datacenter #3, Host #0

0.2: Broker2: Sending Cloudlet 0 to VM #0

160.1: Broker1: Cloudlet 0 received

160.1: Broker1: All Cloudlets executed. Finishing...

160.1: Broker1: Destroying VM #0

Broker1 is shutting down...

160.2: Broker2: Cloudlet 0 received

160.2: Broker2: All Cloudlets executed. Finishing...

160.2: Broker2: Destroying VM #0

Broker2 is shutting down...

Simulation: No more future events

CloudInformationService: Notify all CloudSim entities for shutting down.

Datacenter_0 is shutting down...

Datacenter_1 is shutting down...

Broker1 is shutting down...

Broker2 is shutting down...

Simulation completed.

TABLE A.5
Output: Example 5

Cloudlet ID	STATUS	Data Center ID	VM ID	Time	Start Time	Finish Time
0	SUCCESS	2	0	160	0.1	160.1

TABLE A.6
Output: User 5:Output

Cloudlet ID	STATUS	Data Center ID	VM ID	Time	Start Time	Finish Time
0	SUCCESS	3	0	160	0.2	160.2

A.2.6 CloudSim Example 6: An example showing how to create scalable simulations

Algorithm 6: To create scalable simulations
Inputs: Two VMs id, VM Size, PEs configurations
Output: User Cloudlets
Steps:
Step 1: Creates a container to store VMs. This list is passed to the broker later
Step 2: Create VMs with a space shared scheduling policy for Cloudlets
Step 4: Creates a container to store Cloudlets
Step 5: Cloudlet parameters to be considered: length, fileSize, outputSize;
Step 6: Initialize the CloudSim package. It should be called before creating any entities
Step 7: Initialize the CloudSim library
Step 8: Create Datacenters
Step 9: Create Broker
Step 10: Create VMs and Cloudlets and send them to broker
Step 11: Start the simulation
Step 12: Print results when simulation is over
Step 13: Create PEs and add these into the list
Step 16: Create Hosts with its id and list of PEs and add them to the list of machines
Step 17: Create a DatacenterCharacteristics object that stores the properties of a data center: architecture, OS, list of Machines, allocation policy: time- or space-shared, time zone and its price (G$/Pe time unit)
Step 18: Create a PowerDatacenter object
Step 19: Prints the Cloudlet objects

SOURCE CODE

public class CloudSimExample6
/** The Cloudlet list. */

private static List< *Cloudlet* > CloudletList;

```
/** The vmlist. */
private static List< Vm > vmlist;
private static List< Vm > createVM(int userId, int vms)
```

//Creates a container to store VMs. This list is passed to the broker later
```
LinkedList< Vm > list = new LinkedList< Vm >();
```

//VM Parameters
```
long size = 10000; //image size (MB)
int ram = 512; //vm memory (MB)
int mips = 1000;
long bw = 1000;
int pesNumber = 1; //number of cpus
String vmm = "Xen"; //VMM name
```

//create VMs
```
Vm[] vm = new Vm[vms];
```

```
for(int i=0;i< vms;i++)
vm[i] = new Vm(i, userId, mips, pesNumber, ram, bw, size, vmm, new
CloudletSchedulerTimeShared());
```

//for creating a VM with a space shared scheduling policy for Cloudlets:
```
//vm[i] = Vm(i, userId, mips, pesNumber, ram, bw, size, priority, vmm, new
CloudletSchedulerSpaceShared());
list.add(vm[i]);
```

```
return list;
```

```
private static List< Cloudlet > createCloudlet(int userId, int Cloudlets)
```

// Creates a container to store Cloudlets

```
LinkedList< Cloudlet > list = new LinkedList< Cloudlet >();
```

//Cloudlet parameters
```
long length = 1000;
long fileSize = 300;
long outputSize = 300;
int pesNumber = 1;
UtilizationModel utilizationModel = new UtilizationModelFull();
```

```
Cloudlet[] Cloudlet = new Cloudlet[Cloudlets];
for(int i=0;i<Cloudlets;i++)
Cloudlet[i] = new Cloudlet(i, length, pesNumber, fileSize, outputSize, utiliza-
tionModel, utilizationModel,
utilizationModel);
// setting the owner of these Cloudlets
Cloudlet[i].setUserId(userId);
list.add(Cloudlet[i]);

return list;


/**
 * Creates main() to run this example
 */
public static void main(String[] args)
Log.printLine("Starting CloudSimExample6...");

try
// First step: Initialize the CloudSim package. It should be called
// before creating any entities.
int num_user = 1; // number of grid users
Calendar calendar = Calendar.getInstance();
boolean trace_flag = false; // mean trace events

// Initialize the CloudSim library

CloudSim.init(num_user, calendar, trace_flag);

// Second step: Create Datacenters
//Datacenters are the resource providers in CloudSim. We need at list one of
them to run a CloudSim simulation
Datacenter datacenter0 = createDatacenter("Datacenter_0");
Datacenter datacenter1 = createDatacenter("Datacenter_1");

//Third step: Create Broker
DatacenterBroker broker = createBroker();
int brokerId = broker.getId();

//Fourth step: Create VMs and Cloudlets and send them to broker
vmlist = createVM(brokerId,20); //creating 20 vms
CloudletList = createCloudlet(brokerId,40); // creating 40 Cloudlets

broker.submitVmList(vmlist);
broker.submitCloudletList(CloudletList);
```

```
// Fifth step: Starts the simulation
CloudSim.startSimulation();

// Final step: Print results when simulation is over
List< Cloudlet > newList = broker.getCloudletReceivedList();
CloudSim.stopSimulation();

printCloudletList(newList);
Log.printLine("CloudSimExample6 finished!");

catch (Exception e)

e.printStackTrace();
Log.printLine("The simulation has been terminated due to an unexpected er-
ror");

private static Datacenter createDatacenter(String name)

// Here are the steps needed to create a PowerDatacenter:
// 1. We need to create a list to store one or more
// Machines
List< Host > hostList = new ArrayList< Host >();

// 2. A Machine contains one or more PEs or CPUs/Cores. Therefore, should
// create a list to store these PEs before creating
// a Machine.
List< Pe > peList1 = new ArrayList< Pe >();

int mips = 1000;

// 3. Create PEs and add these into the list.
//for a quad-core machine, a list of 4 PEs is required:

peList1.add(new Pe(0, new PeProvisionerSimple(mips))); // need to store Pe
id and MIPS Rating
peList1.add(new Pe(1, new PeProvisionerSimple(mips)));
peList1.add(new Pe(2, new PeProvisionerSimple(mips)));
peList1.add(new Pe(3, new PeProvisionerSimple(mips)));

//Another list, for a dual-core machine
List< Pe > peList2 = new ArrayList< Pe >();

peList2.add(new Pe(0, new PeProvisionerSimple(mips)));
peList2.add(new Pe(1, new PeProvisionerSimple(mips)));
```

//4. Create Hosts with its id and list of PEs and add them to the list of machines
int hostId=0;
int ram = 2048; //host memory (MB)
long storage = 1000000; //host storage
int bw = 10000;

hostList.add(
new Host(
hostId,
new RamProvisionerSimple(ram),
new BwProvisionerSimple(bw),
storage,
peList1,
new VmSchedulerTimeShared(peList1)
)
); // This is our first machine

hostId++;

hostList.add(

new Host(
hostId,
new RamProvisionerSimple(ram),
new BwProvisionerSimple(bw),
storage,
peList2,
new VmSchedulerTimeShared(peList2)
)
); // Second machine

// 5. Create a DatacenterCharacteristics object that stores the
// properties of a data center: architecture, OS, list of
// Machines, allocation policy: time- or space-shared, time zone
// and its price (G$/Pe time unit).
String arch = "x86"; // system architecture
String os = "Linux"; // operating system
String vmm = "Xen";
double time_zone = 10.0; // time zone this resource located
double cost = 3.0; // the cost of using processing in this resource
double costPerMem = 0.05; // the cost of using memory in this resource
double costPerStorage = 0.1; // the cost of using storage in this resource

```
double costPerBw = 0.1; // the cost of using bw in this resource
LinkedList< Storage > storageList = new LinkedList< Storage >(); //we
are not adding SAN devices by now

DatacenterCharacteristics characteristics = new DatacenterCharacteristics(
arch, os, vmm, hostList, time_zone, cost, costPerMem, costPerStorage, cost-
PerBw);

// 6. Finally, we need to create a PowerDatacenter object.
Datacenter datacenter = null;
try
igdatacenter = new Datacenter(name, characteristics, new VmAllocationPol-
icySimple(hostList), storageList, 0);
catch (Exception e)
e.printStackTrace();

return datacenter;

private static DatacenterBroker createBroker()
DatacenterBroker broker = null;
try
broker = new DatacenterBroker("Broker");
catch (Exception e)
e.printStackTrace();
return null;

return broker;

/* Prints the Cloudlet objects */
private static void printCloudletList(List< Cloudlet > list)
int size = list.size();
Cloudlet Cloudlet;
Log.printLine("=========== OUTPUT ===========");
Log.printLine("Cloudlet ID" + "STATUS" + "Data center ID" + "VM ID"
+ "Time" + "Start Time" + "Finish Time");
for (int i = 0; i < size; i++)
Cloudlet = list.get(i);
Log.print(Cloudlet.getCloudletId());

if (Cloudlet.getCloudletStatus() == Cloudlet.SUCCESS)
Log.print("SUCCESS");

Log.printLine( Cloudlet.getResourceId() + Cloudlet.getVmId() + (Cloudlet.
getActualCPUTime() +
Cloudlet.getExecStartTime()+ Cloudlet.getFinishTime()));
```

Output:CloudSim Example6 Starting CloudSimExample6...
Initialising...
Starting CloudSim version 3.0
Datacenter_0 is starting...
Datacenter_1 is starting...
Broker is starting...
Entities started.
0.0: Broker: Cloud Resource List received with 2 resource(s)
...0.0: Broker: Trying to Create VM #3 in Datacenter_0
......0.0: Broker: Trying to Create VM #17 in Datacenter_0
0.0: Broker: Trying to Create VM #18 in Datacenter_0
0.0: Broker: Trying to Create VM #19 in Datacenter_0
[VmScheduler.vmCreate] Allocation of VM #6 to Host #0 failed by RAM
......[VmScheduler.vmCreate] Allocation of VM #18 to Host #1 failed by
MIPS
[VmScheduler.vmCreate] Allocation of VM #19 to Host #0 failed by RAM
[VmScheduler.vmCreate] Allocation of VM #19 to Host #1 failed by MIPS
0.1: Broker: VM #0 has been created in Datacenter #2, Host #0
...0.1: Broker: VM #5 has been created in Datacenter #2, Host #1
0.1: Broker: Creation of VM #6 failed in Datacenter #2
...0.1: Broker: Creation of VM #19 failed in Datacenter #2
...0.1: Broker: Trying to Create VM #13 in Datacenter_1
0.1: Broker: Trying to Create VM #14 in Datacenter_1

–

–

–

[VmScheduler.vmCreate] Allocation of VM #12 to Host #0 failed by RAM
......
Broker is shutting down...
Simulation: No more future events
CloudInformationService: Notify all CloudSim entities for shutting down.
Datacenter_0 is shutting down...
Datacenter_1 is shutting down...
Broker is shutting down...
Simulation completed.

A.2.7 CloudSim Example 7: An initial example on the use of container simulation.

Algorithm 7: An initial example on the use of container simulation
Inputs:
Multiple VMs id, VM Size, PEs configurations
Output: Cloudlets List
Description: Step 1: Initialize the CloudSim Package before creating entities
Step 2: Define the container allocation Policy. This policy determines how Containers are allocated to VMs in the data center
Step 3: Define the VM selection Policy. This policy determines which VMs should be selected for migration when a host is identified as over-loaded
Step 4: Define the host selection Policy. This policy determines which hosts should be selected as migration destination
Step 5: Define the thresholds for selecting the under-utilized and over-utilized hosts
Step 6: The host list is created considering the number of hosts, and host types
Step 7: Define container allocation policy which defines the allocation of VMs to containers
Step 8: Define the overbooking factor for allocating containers to VMs. This factor is used by the broker for the creating the cloudlet, container and VM lists for submitting to the broker
Step 9: Assign the address for logging the statistics of the VMs, containers in the data center
Step 10: Submit the cloudlet's, container's, and VM's lists to the broker
Step 11: Determining the simulation termination time according to the cloudlet's workload
Step 12: Create a specific name for the experiment which is used for creating the Log address folder
Step 13: Print the Cloudlet objects
Step 14: Create the Virtual machines and add them to the list
Step 15: Create the host list considering the specs
Step 16: Create the data center
Step 17: Create the containers for hosting the cloudlets and binding them together
Step 18: Creating the cloudlet list that are going to run on containers
Step 19: Print the results when the simulation is finished

SOURCE CODE

public class CloudSimExample7

```
{
private static List<Cloudlet> CloudletList;
private static List< Vm > vmlist;
public static void main(String[] args)
{
Log.printLine("Starting CloudSimExample7...");
try {
/* First step: Initialize the CloudSim package. It should be called before cre-
ating any entities. */
int num_user = 1; // number of Cloud users
Calendar calendar = Calendar.getInstance();
boolean trace_flag = false; // trace events
CloudSim.init(num_user, calendar, trace_flag);

/* Defining the container allocation Policy. This policy determines how Con-
tainers are
allocated to VMs in the data center. */

ContainerAllocationPolicy containerAllocationPolicy = new PowerContainer-
AllocationPolicySimple();

/* Defining the VM selection Policy. This policy determines which VMs should
be selected for migration
when a host is identified as over-loaded.*/

PowerContainerVmSelectionPolicy vmSelectionPolicy = new PowerContain-
erVmSelectionPolicyMaximumUsage();

/* Defining the host selection Policy. This policy determines which hosts
should be selected as migration destination
*
*/

HostSelectionPolicy hostSelectionPolicy = new HostSelectionPolicyFirstFit();

/**
* Defining the thresholds for selecting the under-utilized and over-utilized
hosts.
*/

double overUtilizationThreshold = 0.80;
double underUtilizationThreshold = 0.70;
```

/** Prints the Cloudlet objects

private static void printCloudletList(List< *Cloudlet* > list)
int size = list.size();
Cloudlet Cloudlet;
Log.printLine("=========== OUTPUT ===========");
Log.printLine("Cloudlet ID" + "STATUS" + "Data center ID" + "VM ID"
+ "Time" + "Start Time" + "Finish Time");
for (int i = 0; i < size; i++)
Cloudlet = list.get(i);
Log.print(+ Cloudlet.getCloudletId());

if (Cloudlet.getCloudletStatus() == Cloudlet.SUCCESS)
Log.print("SUCCESS");

Log.printLine(Cloudlet.getResourceId() + Cloudlet.getVmId() + (Cloudlet.
getActualCPUTime()) +
Cloudlet.getExecStartTime())+ + (Cloudlet.getFinishTime());

TABLE A.7
Output

Cloudlet ID	STATUS	Data Center ID	VM ID	Time	Start Time	Finish Time
4	SUCCESS	2	4	3	0.2	3.2
...30	SUCCESS	3	6	3	0.2	3
-						
-						
38	SUCCESS	2	2	4	0.2	4.2
...39	SUCCESS	2	3	4	0.2	4.2

Output: CloudSim Example7

Starting CloudSimExample7...

Initialising...

Starting CloudSim version 3.0

Datacenter_0 is starting...

Datacenter_1 is starting...

Broker_0 is starting...

Entities started.

0.0: Broker_0: Cloud Resource List received with 2 resource(s)

0.0: Broker_0: Trying to Create VM #0 in Datacenter_0

...0.1: Broker_0: VM #0 has been created in Datacenter #2, Host #0

...0.1: Broker_0: VM #4 has been created in Datacenter #2, Host #0

0.1: Broker_0: Sending Cloudlet 0 to VM #0

...0.1: Broker_0: Sending Cloudlet 9 to VM #4

200.0: The simulation is paused for 5 sec

Adding: Broker_1

Broker_1 is starting...

200.0: Broker_1: Cloud Resource List received with 2 resource(s)

...200.0: Broker_1: Trying to Create VM #104 in Datacenter_0

200.1: Broker_1: VM #100 has been created in Datacenter #2, Host #1

...200.1: Broker_1: Sending Cloudlet 100 to VM #100

200.1: Broker_1: Sending Cloudlet 101 to VM #101

............519.996: Broker_1: Destroying VM #104

Broker_1 is shutting down...

Simulation: No more future events

CloudInformationService: Notify all CloudSim entities for shutting down.

Datacenter_0 is shutting down...

......Simulation completed.

TABLE A.8

Output: Example 7

Cloudlet ID	STATUS	Data center ID	VM ID	Time	Start Time	Finish Time
0	SUCCESS	2	0	320	0.1	320.1
...3	SUCCESS	2	3	320	0.1	320.1
8	SUCCESS	2	3	320	0.1	320.1

A.2.8 **CloudSim Example 8: An example showing how to create simulation entities (a DatacenterBroker in this example) in run-time using a global manager entity (GlobalBroker)**

Algorithm 8: Creation of simulation entities (a DatacenterBroker in this example) in run-time using a global manager entity (GlobalBroker)
Inputs:
Multiple VMs id, VM Size, PEs configurations
Output: Cloudlets List
Description:
Step 1: Initialize the CloudSim Package before creating entities
Step 2: Define the container allocation Policy. This policy determines how Containers are allocated to VMs in the data center
Step 3: Define the VM selection Policy. This policy determines which VMs should be selected for migration when a host is identified as over-loaded
Step 4: Define the host selection Policy. This policy determines which hosts should be selected as migration destination
Step 5: Define the thresholds for selecting the under-utilized and over-utilized hosts
Step 6: The host list is created considering the number of hosts, and host types
Step 7: Define container allocation policy which defines the allocation of VMs to containers
Step 8: Define the overbooking factor for allocating containers to VMs. This factor is used by the broker for the creating the cloudlet, container and VM lists for submitting to the broker
Step 9: Assign the address for logging the statistics of the VMs, containers in the data center
Step 10: Submit the cloudlet's, container's and VM's lists to the broker
Step 11: Determining the simulation termination time according to the cloudlet's workload
Step 12: Create a specific name for the experiment which is used for creating the Log address folder
Step 13: Print the Cloudlet objects
Step 14: Create the Virtual machines and add them to the list
Step 15: Create the host list considering the specs
Step 16: Create the data center
Step 17: Create the containers for hosting the cloudlets and binding them together
Step 18: Creating the cloudlet list that are going to run on containers
Step 19: Print the results when the simulation is finished

SOURCE CODE

```
public class CloudSimExample8

/** The Cloudlet list. */

private static List< Cloudlet > CloudletList;

/** The vmList. */

private static List< Vm > vmList;

    private static List< Vm > createVM(int userId, int vms, int idShift)

LinkedList< Vm > list = new LinkedList< Vm >();

//VM Parameters

long size = 10000; //image size (MB)
int ram = 512; //vm memory (MB)
int mips = 250;
long bw = 1000;
int pesNumber = 1; //number of cpus
String vmm = "Xen"; //VMM name

//create VMs

Vm[] vm = new Vm[vms];

    for(int i=0;i< vms;i++)
vm[i] = new Vm(idShift + i, userId, mips, pesNumber, ram, bw, size, vmm,
new CloudletSchedulerTimeShared());
list.add(vm[i]);


    return list;


    private static List< Cloudlet > createCloudlet(int userId, int Cloudlets,
int idShift)


LinkedList< Cloudlet > list = new LinkedList< Cloudlet >();

//Cloudlet parameters
```

```
long length = 40000;
long fileSize = 300;
long outputSize = 300;
int pesNumber = 1;
UtilizationModel utilizationModel = new UtilizationModelFull();

    Cloudlet[] Cloudlet = new Cloudlet[Cloudlets];

for(int i=0;i<Cloudlets;i++)

Cloudlet[i] = new Cloudlet(idShift + i, length, pesNumber, fileSize, output-
Size, utilizationModel, utilizationModel, utilizationModel);

Cloudlet[i].setUserId(userId);
list.add(Cloudlet[i]);

    return list;

    public static void main(String[] args)


Log.printLine("Starting CloudSimExample8...");


    try
int num_user = 2; // number of grid users

Calendar calendar = Calendar.getInstance();

boolean trace_flag = false; // mean trace events

CloudSim.init(num_user, calendar, trace_flag);

    GlobalBroker globalBroker = new GlobalBroker("GlobalBroker");

Datacenter datacenter0 = createDatacenter("Datacenter_0");
Datacenter datacenter1 = createDatacenter("Datacenter_1");

DatacenterBroker broker = createBroker("Broker_0");
int brokerId = broker.getId();
vmList = createVM(brokerId, 5, 0); //creating 5 vms
CloudletList = createCloudlet(brokerId, 10, 0); // creating 10 Cloudlets

    broker.submitVmList(vmList);
```

broker.submitCloudletList(CloudletList);

// Fifth step: Starts the simulation
CloudSim.startSimulation();
// Final step: Print results when simulation is over

List< *Cloudlet* > newList = broker.getCloudletReceivedList();
newList.addAll(globalBroker.getBroker().getCloudletReceivedList());
CloudSim.stopSimulation();
printCloudletList(newList);
Log.printLine("CloudSimExample8 finished!");

catch (Exception e)

e.printStackTrace();
Log.printLine("The simulation has been terminated due to an unexpected error");

private static Datacenter createDatacenter(String name)

// Here are the steps needed to create a PowerDatacenter:
// 1. We need to create a list to store one or more
// Machines

List< *Host* > hostList = new ArrayList< *Host* >();

// 2. A Machine contains one or more PEs or CPUs/Cores. Therefore, should
// create a list to store these PEs before creating
// a Machine.

List< *Pe* > peList1 = new ArrayList< *Pe* >(); int mips = 1000;
peList1.add(new Pe(0, new PeProvisionerSimple(mips))); // need to store Pe id and MIPS Rating
peList1.add(new Pe(1, new PeProvisionerSimple(mips)));
peList1.add(new Pe(2, new PeProvisionerSimple(mips)));
peList1.add(new Pe(3, new PeProvisionerSimple(mips)));

//Another list, for a dual-core machine
List< *Pe* > peList2 = new ArrayList< *Pe* >(); peList2.add(new Pe(0, new PeProvisionerSimple(mips))); peList2.add(new Pe(1, new PeProvisionerSimple(mips)));

//4. Create Hosts with its id and list of PEs and add them to the list of machines

```
int hostId=0; int ram = 16384; //host memory (MB)
long storage = 1000000; //host storage
int bw = 10000;

hostList.add(

new Host(
hostId,
new RamProvisionerSimple(ram),
new BwProvisionerSimple(bw),
storage,
peList1,
new VmSchedulerTimeShared(peList1)
)
); // This is our first machine

    hostId++;

    hostList.add(
new Host(
hostId,
new RamProvisionerSimple(ram),
new BwProvisionerSimple(bw),
storage,
peList2,
new VmSchedulerTimeShared(peList2)
)
); // Second machine
```

// 5. Create a DatacenterCharacteristics object that stores the
// properties of a data center: architecture, OS, list of
// Machines, allocation policy: time or space-shared, time zone
// and its price (G$/Pe time unit).

```
String arch = "x86"; // system architecture
String os = "Linux"; // operating system
String vmm = "Xen";
double time_zone = 10.0; // time zone this resource located
double cost = 3.0; // the cost of using processing in this resource
double costPerMem = 0.05; // the cost of using memory in this resource
double costPerStorage = 0.1; // the cost of using storage in this resource
double costPerBw = 0.1; // the cost of using bw in this resource
```

LinkedList< *Storage* > storageList = new LinkedList< *Storage* >(); //we are not adding SAN devices by now

DatacenterCharacteristics characteristics = new DatacenterCharacteristics(arch, os, vmm, hostList, time_zone, cost, costPerMem, costPerStorage, cost-PerBw);

    // 6. Finally, we need to create a PowerDatacenter object.

Datacenter datacenter = null;

try
datacenter = new Datacenter(name, characteristics, new VmAllocationPoli-cySimple(hostList), storageList, 0);

catch (Exception e)
e.printStackTrace();

return datacenter;

    private static DatacenterBroker createBroker(String name)

    DatacenterBroker broker = null;
try
broker = new DatacenterBroker(name);
catch (Exception e)
e.printStackTrace();
return null;

return broker;

Output: CloudSimExample8 Starting CloudSimExample8...
Initialising...
Starting CloudSim version 3.0
GlobalBroker is starting...
Datacenter_0 is starting...
Datacenter_1 is starting...
Broker_0 is starting...
Entities started.
0.0: Broker_0: Cloud Resource List received with 2 resource(s)
0.0: Broker_0: Trying to Create VM #0 in Datacenter_0
...0.0: Broker_0: Trying to Create VM #3 in Datacenter_0
0.0: Broker_0: Trying to Create VM #4 in Datacenter_0
0.1: Broker_0: VM #0 has been created in Datacenter #3, Host #0
...0.1: Broker_0: VM #3 has been created in Datacenter #3, Host #1
0.1: Broker_0: VM #4 has been created in Datacenter #3, Host #0
0.1: Broker_0: Sending Cloudlet 0 to VM #0
......0.1: Broker_0: Sending Cloudlet 9 to VM #4
Adding: GlobalBroker_ GlobalBroker_ is starting...
200.0: GlobalBroker_: Cloud Resource List received with 2 resource(s)
200.0: GlobalBroker_: Trying to Create VM #100 in Datacenter_0
...200.1: GlobalBroker_: VM #100 has been created in Datacenter #3, Host #1
200.1: GlobalBroker_: VM #101 has been created in Datacenter #3, Host #0
200.1: GlobalBroker_: VM #102 has been created in Datacenter #3, Host #1
...320.1: Broker_0: Cloudlet 0 received
320.1: Broker_0: Cloudlet 5 received
...320.1: Broker_0: Destroying VM #0
...320.1: Broker_0: Destroying VM #3
320.1: Broker_0: Destroying VM #4
Broker_0 is shutting down...
520.1: GlobalBroker_: Cloudlet 101 received
520.1: GlobalBroker_: Cloudlet 106 received
...520.1: GlobalBroker_: Cloudlet 109 received
...520.1: GlobalBroker_: Destroying VM #103
...Simulation completed.

TABLE A.9
Output: Example8

Cloudlet ID	STATUS	Data center ID	VM ID	Time	Start Time	Finish Time
0	SUCCESS	3	0	320	0.1	320.1
...8	SUCCESS	3	3	320	0.1	320.1
...109	SUCCESS	3	104	320	200.1	520.1

Bibliography

[1] Google App Engine, *Available at: http://appengine.google.com*, Accessed on 18 April 2020.

[2] Quiroz A, Kim H, Parashar M, Gnanasambandam N, Sharma N., "Towards autonomic workload provisioning for enterprise grids and Clouds", *Proceedings of the 10th IEEE/ACM International Conference on Grid Computing*, Banf, AB, Canada, 2009, pp. 50-57.

[3] Dumitrescu CL, Foster I., "GangSim: A simulator for grid scheduling studies", *Proceedings of the IEEE International Symposium on Cluster Computing and the Grid*, Cardiff, U.K., 2015, pp. 1151-1158.

[4] Legrand A, Marchal L, Casanova II., "Scheduling distributed applications: The SimGrid simulation framework", *Proceedings of the Third IEEE/ACM International Symposium on Cluster Computing and the Grid*, Tokyo, Japan, 2003, pp. 138-145.

[5] Buyya R, Murshed M., "GridSim: A toolkit for the modeling and simulation of distributed resource management and scheduling for grid computing", *Concurrency and Computation Practice and Experience*, Vol. 14, No. 13, 2007, pp. 1175-1220.

B

Appendix B: Experiments Using Cloud Platforms

Following experiments are covered in this chapter:

1. Illustration of Backup-Restore for VMs

2. Illustration of the VMs Cloning

3. Evaluation of the performance of MapReduce Program on Word Count for different file size

4. Provisioning Communication between Multiple VMs with and without vClient on a single physical machine

5. Installation and configuration of virtualization using Kernal Virtual Machine (KVM)

6. Study and implementation of IaaS

7. Study and implementation of Storage-as-a-Service (StaaS)

8. Study and implementation of Identity management

9. Study of Cloud Security management

10. Working and installation of Google App Engine

11. Database Stored Procedure using Microsoft Azure

B.1   Installation of Platforms

Here, we discuss about installation of Tools or Platforms, for which the experiments are covered in this chapter.

VMware Workstation

1. Login into server as root or non-root user with sudo permissions.
   # yum update   [On RedHat Systems]

   # dnf update   [On Fedora]

2. Download the VMWare Workstation Pro installer script bundle from VMware official site.

3. After downloading the VMWare Workstation Proscript file, go to the directory which contains the script file and set the appropriate execute permission as shown.
   # chmod a+x VMware-Workstation-Full-15.5.1-15018445.x86_64.bundle

4. Installing VMWare Workstation 15 Pro in Linux

5. Now run the installer script to install VMWare Workstation Pro on a Linux host system, with the installation progress shown in the terminal.
   # ./VMware-Workstation-Full-15.5.1-15018445.x86_64.bundle OR
   $ sudo ./VMware-Workstation-Full-15.5.1-15018445.x86_64.bundle

6. Running VMWare Workstation 15 Pro

7. To start the software, type vmware in the terminal.
   [root@tecmint Downloads]# vmware

   After running the above command, if you do not have GCC GNU C Compiler, you will see the message which notifies you to install the GCC compiler and some components. Just press "Cancel" to continue.

8. Return to the terminal to install "Development Tools"

[root@tecmint Downloads]# yum groupinstall "Development tools" [On RedHat Systems]

root@tecmint: # apt-get install build-essential [On Debian Systems]

[root@tecmint Downloads]# vmware

[root@tecmint Downloads]# rpm -qa — grep kernel-headers [On RedHat systems]

root@tecmint: # dpkg -l — grep linux-headers [On Debian systems]

[root@tecmint]# yum install kernel-headers [On RedHat Systems]

root@tecmint: # apt-get install linux-headers-'uname -r' [On Debian Systems]

Hadoop platform

We discuss the set up and configuration of a single-node Hadoop installation so that we can quickly perform simple operations using Hadoop MapReduce and the Hadoop Distributed File System (HDFS). Windows is a supported platform but the followings steps are for Linux only. The pre-requisites is that Java must be installed, "ssh" must be installed and "sshd" must be running to use the Hadoop scripts that manage remote Hadoop daemons.

1. Start your Hadoop cluster in one of the three supported modes: (i) Local (Standalone) Mode, (ii) Pseudo-Distributed Mode and (iii) Fully-Distributed Mode.
2. Check whether you can "ssh" to the localhost without a passphrase.
   $ ssh localhost

3. If you cannot ssh to localhost without a passphrase, execute the following commands:

   $ ssh-keygen -t rsa -P " -f /.ssh/id_rsa
   $ cat /.ssh/id_rsa.pub ¿¿ /.ssh/authorized_keys
   $ chmod 0600 /.ssh/authorized_keys

4. Format the filesystem
   $ bin/hdfs namenode -format

5. Start NameNode daemon and DataNode daemon:
   $ sbin/start-dfs.sh

6. The hadoop daemon log output is written to the $HADOOP_LOG_DIR directory (defaults to $HADOOP_HOME/logs)

7. Browse the web interface for the NameNode; by default it is available at:
   NameNode - http://localhost:50070/

8. Make the HDFS directories required to execute MapReduce jobs:

   $ bin/hdfs dfs -mkdir /user
   $ bin/hdfs dfs -mkdir /user/<username>

Kernel-based Virtual Machine (KVM)

KVM is a full virtualization solution for Linux that is included in the mainline Linux kernel since 2.6.20 and is stable and fast for most workloads. Here we shall install qemu-kvm and qemu-img packages at first.

1. Install the packages provide the user-level KVM and disk image manager

   [root@server]# yum install qemu-kvm qemu-img

2. Tools to administrate virtual platform needs to be installed. These are: (i) virt-manager, a GUI tool to administrate your virtual machines, (ii) libvirt-client, a CL tool to administrate your virtual environment, (iii) virt-install provides the command "virt-install" to create your virtual machines from Command Line Interface (CLI), (iv) libvirt provides the server and host side libraries for interacting with hypervisors and host systems. These tool can be installed using the following command.

   [root@server]# yum install virt-manager libvirt libvirt-python libvirt-client

3. The virtualization daemon which manage all of the platform is "libvirtd". It can be started using the following command.

   [root@server]#systemctl restart libvirtd

4. After restarting the daemon, then check its status by running following command.

   [root@server]#systemctl status libvirtd

5. Select the installation method which you will use to create the virtual machine. For now let us discuss installation on Local install media, so you should provide the path of your "ISO" image.

6. The storage has return back, hence use the virtual disk to install virtual machine on it.

7. The final step asks about the name of the virtual machine and another advanced options. Click "finish", after which control console will appear for your Guest OS to be managed.

OwnCloud on Ubuntu

OwnCloud is a leading open-source file sharing and Cloud collaboration platform whose services and functionalities are similar to those offered by DropBox and Google Drive. However, unlike Dropbox, OwnCloud does not have the datacenter capacity to store hosted files. Nevertheless, you can still share files such as documents, images, and videos to mention a few and access them across multiple devices such as smartphones, tablets and PCs.

Here, we learn the installation of OwnCloud on Ubuntu 18.04 and newer versions.

1. Update the system packages and repositories using the following "apt" command.

   $ sudo apt update -y & sudo apt upgrade -y

2. OwnCloud is built on PHP and is typically accessed via a web interface. For this reason, we are going to install Apache web server to serve OwnCloud files as well as PHP 7.2 and additional PHP

modules necessary for OwnCloud to function smoothly. Type the following command:

$ sudo apt install apache2 libapache2-mod-php7.2 openssl php-imagick php7.2-common php7.2-curl php7.2-gd php7.2-imap php7.2-intl php7.2-json php7.2-ldap php7.2-mbstring php7.2-mysql php7.2-pgsql php-smbclient php-ssh2 php7.2-sqlite3 php7.2-xml php7.2-zip

3. Once the installation is complete, you can verify if Apache is installed by running the "dpkg" command

   $ sudo dpkg -l apache

4. To start and enable Apache to run on boot, run the command

   $ sudo systemctl enable apache2

5. Open your browser and type in your server's IP address in the URL bar by following command

   http://server-IP

6. You should get a webpage showing that Apache is installed and running. To check if PHP is installed, type the following

   $ php -v

7. Install MariaDB in Ubuntu. To install MariaDB run the following command

   $ sudo apt install mariadb-server

8. To get started with securing your MySQL server, run the command

   $ sudo mysql_secure_installation

9. Hit ENTER when prompted for the root password and press "Y" to set the root password. For the remaining prompts, simply type "Y" and hit ENTER.

10. Create a "OwnCloud" Database. We need to create a database for OwnCloud to store files during and after installation. So log in to MariaDB

    $ sudo mysql -u root -p

11. Run the commands given as under:

    MariaDB [(none)]> CREATE DATABASE ownCloud_db;
    MariaDB [(none)]> GRANT ALL ON ownCloud_db.* TO 'own-Cloud_user'@localhost';
    MariaDB [(none)]> FLUSH PRIVILEGES;
    MariaDB [(none)]> EXIT;

12. Create OwnCloud Database in Ubuntu.

13. Download OwnCloud in Ubuntu.

14. After creating the database, now download the OwnCloud zipped file using the following wget command

    $ sudo wget https://download.ownCloud.org/community/ownCloud-10.4.0.zip

15. Once downloaded, unzip the zipped package to the /var/www/ directory

    $ sudo unzip ownCloud-10.4.0.zip -d /var/www/

16. Then, set permissions.

    $ sudo chown -R www-data:www-data /var/www/ownCloud/
    $ sudo chmod -R 755 /var/www/ownCloud/

17. Configure Apache for OwnCloud

    $ sudo vim /etc/apache2/conf-available/ownCloud.conf

18. For the changes to come into effect restart Apache web server

    $ sudo systemctl restart apache2

19. Go to your browser and type in your server's address followed by
     the /ownCloud suffix

AWS Cloud Development Kit (AWS CDK)

The AWS Cloud Development Kit (AWS CDK) is an open source software development framework to model and provision our Cloud application
resources using familiar programming languages. Provisioning Cloud applications can be a challenging process that requires us to perform manual actions,
write custom scripts, maintain templates, or learn domain-specific languages.
AWS CDK uses the familiarity and expressive power of programming languages for modeling our applications. It provides us with high-level components that preconfigure Cloud resources with proven defaults, so that we can
build Cloud applications.

We need to provide our credentials and an AWS Region to use AWS CDK.
Using AWS root account for day-to-day tasks is not recommended. Instead,
creation of a user in Identity and Access Management (IAM) and its usage
with CDK credentials is advised. If AWS CLI is installed, the easiest way
to satisfy this requirement is to install the AWS CLI and issue the following
command:

aws configure

As a next step, we require to provide our AWS access key ID, secret access key, and default region when prompted. We may also manually create or
edit the /.aws/config and /.aws/credentials (Linux or Mac) or %USERPRO-
FILE%àws/config and %USERPROFILE%àws/credentials (Windows) files to
contain credentials and a default region, in the following format.

In /.aws/config or %USERPROFILE%àws/config
[default]
region=us-west-2

In /.aws/credentials or %USERPROFILE%àws/credentials
aws_access_key_id=AKIAI44QH8DHBEXAMPLE
aws_secret_access_key=je7MtGbClwBF/2Zp9Utk/h3yCo8nvbEXAMPLEKEY

Install the AWS CDK Toolkit globally using the following Node Package
Manager command.

npm install -g aws-cdk

Windows Azure SDK

The Azure SDKs are collections of libraries built to make it easier to use Azure services from your language of choice. These libraries are designed to be consistent, approachable, diagnosable, dependable and idiomatic.

Microsoft Azure, the Cloud computing platform is transforming the business use of technology. The main reason that could be attributed is that it supports varied operating systems, tools, databases, programming languages, and devices.

Installation of Windows Azure SDK

1. Go to official site of Windows Azure at link http://www.windowsazure.com/en-us/

2. In the bottom of page, we get Develop option. Click on "Show Me More"

3. Now we need to choose the language we want to work with. We can develop application in any of the language given in option and deploy it on the Microsoft managed datacenters

4. To start developing using .NET, click on .NET

5. We will be navigated to Home Page of .NET Developer Center. We get all the resources related to .NET development on Windows Azure

6. Now click on Install for the installation of Windows Azure SDK.

B.2   Illustration of Backup-Restore for VMs

Aim: To illustrate backup-restore of VMs

Outcomes: The learners will be able to:

1. Learn backup-restore sessions in VM

2. Apply backup-restore sessions for efficient usage of resource in VMs.

Theoretical Description

When we take a snapshot, we capture the state of the VM settings and the virtual disk. If we take a memory snapshot, we also capture the memory state of the VM. These states are saved to files that reside with the VM's base files.

Requirements

Following are the requirements for successful execution of this experiment:

1. VM is powered off taking a memory snapshot of a VM that has multiple disks in different disk modes

2. To quiesce the VM files, verify that the VM is powered on and that VMware is installed

3. Privilege to be set for VM is: Virtual machine.State.

4. A snapshot of the VM should be successfully created

5. The VM to which the backup is required should be in ON state

Procedure

1. Right click on VM − > Snapshot − > Take Snapshot

2. Provide the name for snapshot. It may also be based on the backup time

3. Make changes/updates in the current VM

4. For retrieval process, get back to initial state of VM

5. Right click on VM − > Snapshot − > Revert to Current Snapshot

6. Accept for the current modification to be done in the VM

7. Get back to original state of VM

8. If we take too many snapshots for the same VM, we need to revert to required state of VM

9. Right click on VM − > Snapshot − > Snapshot manager

10. We can see the multiple snapshots taken for the VM and also current state of VM

11. We can select any snapshot and revert to respective state of VM

Conclusion: A VMware snapshot is demonstrated to restore a VM to a particular point when a failure or system error occurs.

B.3   Illustration of the VMs Cloning

Aim: To illustrate cloning of VMs

Outcomes: The learners will be able to:

1. Understand the concepts of cloning.

2. Configure multiple VMs on multiple ESX integrated (ESXi)s.

Theoretical Description

Cloning a VM creates a duplicate of the VM with the same configuration and installed software as the original.

Clones are useful when you must deploy many identical virtual machines to a group.

Requirements

1. To customize the guest operating system of the VM, we need to check that our guest operating system meets the requirements for customization

2. To use a customization specification, we must first create or import the customization specification

3. To use a custom script to generate the host name or IP address for the new VM, we need to configure the script.

Procedure

There are two ways to be followed discussed as follows:

Method I: Using Open Virtualization Format (OVF)

1. Go to Edit Settings of the VM

2. Change the CD/DCD drive from "datasore" to "Client Device"

3. Make sure that the VM is turned OFF

4. Select the VM which you want to clone

5. Go to File − > Export − > Export OVF Template

6. Select the destination where you want to store the copy of it

7. For importing a copy to same /different ESXi, Go to file − > Deploy OVF Template

8. Select the source location from where the OVF file is to be deployed

9. Select the exact copy of VM to be deployed among many files.

Method II: Download/Upload a Folder of VM

1. Select the ESXi. Right click on Datastore − > browse datastore

2. Select the folder of VM to be downloaded to physical machine

3. Select the option of "download a file from this data store to local machine"

4. Select the storage location

5. Go to vClient of any ESXi, Datastore − > Upload folder

6. Once the folder of VM is uploaded successfully, then open the folder

7. Right click on the .vms file − > add to inventory

8. Provide the name for the new VM

9. Select the destination ESXi

10. Once it is installed on to inventory, then power "ON" the VM

Conclusion: A demonstration of VM cloning was done.

B.4 Evaluation of the Performance of MapReduce Program on Word Count for Different File Size

Aim: A demonstration of word count example in Hadoop platform

Outcomes: The learners will be able to:

1. Execute MapReduce programs on Hadoop

2. Analysis of the results in Hadoop architecture.

Theoretical Description

   A MapReduce job usually splits the input data-set into independent chunks which are processed by the map tasks in a completely parallel manner.

Requirements

1. Ubuntu Linux 10.04 LTS (deprecated: 8.10 LTS, 8.04, 7.10, 7.04)

2. Hadoop 1.0.3, released May 2012

3. Hadoop requires a working Java 1.5+ (aka Java 5) installation. However, using Java 1.6 (aka Java 6) is recommended for running Hadoop

Procedure

1. Download example input data. Download each ebook as text files in plain-text UTF-8 encoding and store the files in a local temporary directory of choice, for example /tmp/gutenberg. We use three ebooks from Project Gutenberg for this experiment. They are listed as follows:

    *   The Outline of Science by J. Arthur Thomson

    *   The Notebooks of Leonardo Da Vinci

    *   Ulysses by James Joyce

2. Restart the Hadoop cluster

3. Copy local example data to HDFS

4. Before we run the actual MapReduce job, we first have to copy the files from our local file system to HDFS

5. Run the MapReduce job

6. Check if the result is successfully stored in HDFS directory /user/hduser/gutenberg-output.

Conclusion
The experiment efficiently reduce the burden of HDFS and improve the read and write performance of moderate size files.

B.5 Provisioning Communication between Multiple VMs with and without vClient on a Single Physical Machine

Aim: To provision communication between multiple VMs with and without vClient on a single physical machine

Outcomes: The learner will be able to:

1. To install and setup the VM
2. To check the connectivity between the VMs.

Theoretical Description VMs are systems that abstract over the communication mechanisms of a (potentially heterogeneous) computer cluster. Such a VM does not consist of a single process, but one process per physical machine in the cluster.

Requirements

Like physical computers, the virtual machines running under VMware Workstation generally perform better if they have faster processors and more memory.

1. PC Hardware – Standard x86-compatible personal computer, 400 MHz or faster CPU minimum (500 MHz recommended), Multiprocessor systems supported, 64-bit processor support for AMD64 Opteron, Athlon 64 and Intel IA-32e CPU

2. Memory: 128 MB minimum (256 MB recommended)

3. Disk Drives: IDE and SCSI hard drives supported, up to 950GB capacity. At least 1GB free disk space recommended for each guest operating system and the application software used with it

4. If you use a default setup, the actual disk space needs are approximately the same as those for installing and running the guest operating system and applications on a physical computer.

Procedure:

1. Access the ESXi through the vClient

2. Open vSphere Client and enter the following:

   - IP of the server (eg: 128.0.78.222)
   - Enter the user Name of the server
   - Enter the password of the server

3. To upload to datastore, open the summary tab from the interface

4. Right click on datastore and browse datastore

5. Click on upload icon, press upload file, browse the file and upload

6. To create a VM, follow the procedure

   - Right click on the server > New Virtual Machine
   - Select Custom and click next
   - Give a name to the virtual machine and click next
   - Select storage from the list and click next
   - Select the virtual machine version (Select the latest version) and click next
   - Select the type of guest Operating System and click next

- Select the number of virtual cores (Virtual sockets)
- Select the amount of RAM and click next
- Do not change anything in the networks page and then click next
- Select SCSI controller as LSI Logic parallel
- Click on "create a new disk"
- Enter the disk capacity (Atleast 30GB), select "Thin provisioning" and click next
- Review the configuration and finish
- Now select the VM and click on "Edit Virtual machine settings"
- Click on CD/DVD drive 1, select datastore ISO file and locate the ISO on the datastore.

7. Turn on the virtual machine and proceed with normal installation of OS in to that VM

8. Repeat the same to create one more VM with different name/IP on same ESXi, then later open them through different/same clients.

9. Now PING between them using their IP address. As ping <ip of other vm>

Conclusion

We experimented provisioning communication between multiple VMs with and without vClient on a single physical machine. Running multiple virtual machines on one physical machine enhances its resource utility.

B.6   Installation and Configuration of Virtualization Using KVM

Aim: Installation and Configuration of virtualization using KVM

Outcomes: The learners will be able to:

1. Analyze user models and develop user centric interfaces

2. Analyze the local and global impact of computing on individuals, organizations, and society

3. Engage in life-long learning development and higher studies

4. Understand, identify, analyze and design the problem
5. Implement and validate the solution including both hardware and software.

Theoretical Description KVM or (Kernel-based Virtual Machine) is a full virtualization solution for Linux on Intel 64 and AMD 64 hardware that is included in the mainline Linux kernel since 2.6.20 and is stable and fast for most workloads.

Hardware/Software Required: Ubuntu operating system, open source software KVM, Internet.

Procedure:

- #sudo grep -c "svm∥vmx" /proc/cpuinfo

- #sudo apt-get install qemu-kvm libvirt-bin bridge-utils virt-manager

- #sudoadduserrait
  After running this command, log out and log back in as rait

- #sudoadduserraitlibvirtd
  After running this command, log out and log back in as rait

- Open Virtual Machine Manager application and Create Virtual Machine #virt-manager as shown below

- Install windows operating system on virtual machine

Conclusion
Kernel-based virtualization was implemented to see the advantage of hardware assisted virtualization support given by the new generation of Intel and AMD CPU's.

B.7 Study and Implementation of IaaS

Aim: To study and implementation of IaaS

Outcomes: The learners will be able to:

1. Undertand the industry requirements in the domains of database management, programming and networking with limited infrastructure

2. Analyze the local and global impact of computing on individuals, organizations and society

3. Use current techniques, skills and tools necessary for computing practice.

Theoretical Description KVM or (Kernel-based Virtual Machine) is a full virtualization solution for Linux on Intel 64 and AMD 64 hardware that is included in the mainline Linux kernel since 2.6.20 and is stable and fast for most workloads.

Hardware/Software Required: Ubuntu operating system, Virtual machine, WAMP/ZAMP server, any tool or technology can be used for implementation of web application e.g., JAVA, PHP, etc.

Procedure: Installation Steps: (https://docs.openstack.org/devstack/latest/guides/single-machine.html)

Step 1: Add user
useradd -s /bin/bash -d /opt/stack -m stack
apt-get install sudo -y
echo "stack ALL=(ALL) NOPASSWD: ALL"

Step 2: login as stack user
Download DevStack
sudo apt-get install git -y —— sudo yum install -y git
git clone https://git.openstack.org/openstack-dev/devstack
cd devstack

Step 3: Run DevStack
Now to configure stack.sh. DevStack includes a sample in devstack/samples/local.conf. Create local.conf as shown below to do the following:

- Set the RabbitMQ password

- Set the service password. This is used by the OpenStack services (Nova,

Glance, etc) to authenticate with Keystone. local.conf should look something like this:

[[local |localrc]]
FLOATING_RANGE=192.168.1.224/27
FIXED_RANGE=10.11.12.0/24
FIXED_NETWORK_SIZE=256
FLAT_INTERFACE=eth0
ADMIN_PASSWORD=supersecret
DATABASE_PASSWORD=iheartdatabases
RABBIT_PASSWORD=flopsymopsy
SERVICE_PASSWORD=iheartksl

- Run DevStack: ./stack.sh A seemingly endless stream of activity ensues. When completed you will see a summary of stack.sh's work, including the relevant URLs, accounts and passwords to poke at your shiny new Open-Stack

- Using OpenStack: At this point you should be able to access the dashboard from other computers on the local network.

Conclusion:
We installed Ubuntu/ Xen as bare metal hypervisor and implemented it. It provides access to computing resources in a virtual environment. With the help of Infrastructure as a service we could build our own IT platform.

B.8   Study and Implementation of Identity Management

Aim: To study and implement identity management

Outcomes: The learners will be able to:

1. Understand concepts of virtualization and to use Cloud as IaaS
2. Learn the technique and its complexity
3. Understand the importance of this technique from application point of view

Theoretical Description Identity management is the hierarchical procedure for distinguishing, verifying and approving people or individuals to access applications, frameworks or systems with established identities.

Hardware/Software Required: OwnCloud software.

Procedure:

1. Apps Selection Menu: Located in the upper left corner, click the arrow to open a dropdown menu to navigate to your various available apps. An option "Apps Information field," located in the left sidebar provides filters and tasks associated with the selected app. "Application View" is the central field in the ownCloud user interface. This field displays the contents or user features of your selected app.

2. Share the file or folder with a group or other users, and create public shares with hyperlinks. You can also see whom you have shared with already, and revoke shares by clicking the "trash" can icon. If username auto-completion is enabled when you start typing the user or group name, ownCloud will automatically complete it for you. If your administrator has enabled email notifications, you can send an email notification of the new share from the sharing screen.

3. Five share permissions are: (i) "Can share" allows the users you share with to re-share, (ii) "Can edit" allows the users you share with to edit your shared files, and to collaborate using the "Documents" app, (iii) "Create" allows the users you share with to create new files and add them to the share, (iv) "Change" allows uploading a new version of a shared file and replacing it, (v) "Delete" allows the users you share with to delete shared files.

Conclusion: We have studied how to use ownCloud for ensuring identity management of the users. We can create multiple groups and provide privileges to view or modify data as per defined permissions. It also enables simplified look and feel to be used by anyone.

B.9   Study of Cloud Security Management

Aim: To Study Cloud Security Management

Outcomes: The learners will be able to:

1. Understand the security features of Cloud

2. Learn the technique of application security management and its complexity

3. Analyze Cloud security management from application point of view

Theoretical Description Cloud computing security or, more simply, Cloud security refers to a broad set of policies, technologies, applications and controls utilized to protect virtualized IP, data, applications, services and the associated infrastructure of cloud computing. It is a sub-domain of computer security, network security, and, more broadly, information security.

Requirements Hardware/Software Required: Ubuntu operating system, Virtual machine, WAMP/ZAMP server, any tool or technology for implementation of web application e.g., JAVA, PHP, etc.

Procedure

1.
   - Goto aws.amazon.com
   - Click on "My Account"
   - Select "AWS management console"
   - Give email id in the required field
2. If you are registering first time, then select "I am a new user" option
   - Click on "sign in using our secure server" button
   - Follow the instruction and complete the form filling
   - Sign out from aws.amazon.com
3. Do not provide any credit card details or bank details
   - Again go to "My Account"
   - Select "AWS management console"
   - Sign in again by entering the user name and valid password
   - Check "I am returning user and my password is" radio button
   - Now you are logged in as a "Root User"
4. All AWS project can be viewed by you, but you cannot make any changes in it or you cannot create new thing as you are not paying any charges to amazon. To create the user in a root user, follow the steps mentioned below:

   - Click on "Identity and Access Management" in security and identity project

- Click in "Users" from dashboard"
- Check for "Create New Users"
- Click on create new user button
- Enter the "User Name"
- Select "Generate and access key for each user" checkbox, it will create a user with a specific key
- Click on "Create" button at right bottom
- Once the user is created click on it
- Go to security credentials tab
- Click on "Create Access Key," it will create an access key for user
- Click on "Manage Multi-factor Authentication (MFA) device" it will give you one QR code displayed on the screen
- Scan the QR code on your mobile phone using barcode scanner (install it in mobile phone). Also install "Google Authenticator" in your mobile phone to generate the MFA code

5. Google authenticator will keep on generating a new MFA code after every 60 seconds that code you will have to enter while logging as a user. Hence, the security is maintained by MFA device code

6. One cannot use your AWS account even if it may have your user name and password, because MFA code is on your MFA device (mobile phone in this case) and it is getting changed after every 60 seconds.

7. Permissions in user account: After creating the user by above mentioned steps; you can give certain permissions to specific user

- Click on created user
- Goto "Permissions" tab
- Click on "Attach Policy" button
- Select the needed policy from given list and click on apply.

Conclusion: We have studied how to secure the Cloud and its data. Amazon AWS provides the best security with its extended facilities and services like MFA device. It also gives us the ability to add our own permissions and policies for securing encrypted data.

B.10   Working and Installation of Google App Engine

Aim: Working and installation of Google App Engine

Outcomes: The learners will be able to:

1. Understand important concepts and terminologies for working with Google App Engine

2. Learn the working of Google App Engine.

Theoretical Description Google App Engine (often referred to as GAE or simply App Engine) is a PaaS and Cloud computing platform for developing and hosting web applications in Google-managed data centers. Applications are sandboxed and run across multiple servers. App Engine offers automatic scaling for web applications. As the number of requests increases for an application, App Engine automatically allocates more resources for the web application to handle the additional demand.

Requirements Login account in "codenvy"

Deploying a New Application GAE

Procedure

1. Create a new project from scratch and choose either Java web application or Python and Google App Engine as PaaS (if you have already created a project, then open it and go to PaaS > Google App Engine > Create Application)

2. Enter project name and choose a Template

3. Check Use existing GAE ID if you want to overwrite an existing app

4. Click Create button

5. If you deploy your first app to GAE from Codenvy you need to authenticate

6. Allow access to proceed

7. Enter required information at the GAE webpage

8. Once you click Create Application, the browser's tab will be automatically closed in a few seconds

9. When you are back to your Codenvy workspace, click Deploy to push the app to GAE

10. The process may take several minutes, and you will see a confirmation message in the Output panel with the application url : yourapp-name.appspot.com

11. Make sure you use the same Google ID to log in to Codenvy and Google App Engine.

Import an Existing GAE Application If you have a GAE application which needs to be imported from Codenvy, here are the steps that needs to be followed.

1. Download source code of your app

2. Push the code to GitHub or another remote repository cloning GitHub project to Codenvy

3. Open appengine-web.xml file and edit application ID, if necessary, for example <application>javagae112</application> (enter the app ID you need to update on GAE)

4. If you want to create a new version of the same app, you can change it as well

5. Update application through Project > PaaS > Google App Engine

6. Once the app is updated, it can be changed and updated anytime directly from Codenvy

Conclusion: We studied the steps to be followed in GAE, which is a Platform as a Service and Cloud computing platform for developing and hosting web applications in Google-managed data centers.

B.11 Database Stored Procedure Using Microsoft Azure

Aim: Database Stored Procedure using Microsoft Azure

Outcomes: The learners will be able to:

1. Understand concepts and terminology for working with Microsoft Azure

2. Create a new database stored procedure using the new Azure portal

Theoretical Description A stored procedure is nothing but a group of SQL statements compiled into a single execution plan. A stored procedure is used to retrieve data, modify data and delete data in database table.

Requirements Account on Microsoft Azure

Procedures Let us see how to create a new database stored procedure using the new Azure portal instead of using the SQL Server Management Studio.

1. Login to the portal using your Microsoft Windows Live credentials with Azure credentials

2. In the Database Menu, go to the "Database Subscription Window"

3. To create a new Stored Procedure, click on "Manage" at the top menu tool bar

4. Click on New Stored procedure menu at the top to see a script window for writing customized stored procedure

5. Once we create the structure for the stored procedure, we need to save it. Once saved, we can use the stored procedure to execute the same. We need to navigate to the new query window in the database section and write an execute command

6. We can create "n" number of stored procedure as per the requirement and use it across the process similar to the traditional SQL Server locally.

Conclusion: Stored procedures are created normally using the SQL Serve management studio. With the latest version of SQL Azure, we have option to create a user stored procedure directly online without need to have a local interface. This way we have some control of using it anywhere anytime to do some updates regularly

Bibliography

[1] Google App Engine, *Available at: http://appengine.google.com*, Accessed on 18 April 2020.

[2] Quiroz A, Kim H, Parashar M, Gnanasambandam N, Sharma N., "Towards autonomic workload provisioning for enterprise grids and Clouds", *Proceedings of the 10th IEEE/ACM International Conference on Grid Computing*, Banf, AB, Canada, 2009, pp. 50-57.

[3] Dumitrescu CL, Foster I., "GangSim: A simulator for grid scheduling studies", *Proceedings of the IEEE International Symposium on Cluster Computing and the Grid*, Cardiff, U.K., 2015, pp. 1151-1158.

[4] Legrand A, Marchal L, Casanova H., "Scheduling distributed applications: The SimGrid simulation framework", *Proceedings of the Third IEEE/ACM International Symposium on Cluster Computing and the Grid*, Tokyo, Japan, 2003, pp. 138-145.

[5] Buyya R, Murshed M., "GridSim: A toolkit for the modeling and simulation of distributed resource management and scheduling for grid computing", *Concurrency and Computation Practice and Experience*, Vol. 14, No. 13, 2007, pp. 1175-1220.

Index

ABE, 208
Abstraction, 114
ACO, 162
Admission control, 132
Amazon AWS, 93
Amdahl's Law, 30
Apache CloudStack, 7, 193
API, 104
Application SLA, 148
Archival Storage, 88
ASA, 145
Availabilty, 206
AWS CDK, 196, 304

Binary translation, 66

Capacity management, 130
Capacity planning, 127
CAPEX, 127, 170
CDN, 104
Ceilometer, 191
Chip-multithreading, 48
Cinder, 191
Cloud computing, 2
Cloud controller, 189
Cloud hosting, 5
CloudSim, 245
CloudSim simulator, 245
Cluster computing, 2
Cluster controller, 190
Communication Utility, 161
Compliance, 9
Concurrency, 48
Confidentiality, 205
Consolidation, 68
Containers, 66
Contract Definition, 149

Corrective controls, 223
CP-ABE, 208
CRIS, 163
Customer-based SLA, 144

Data Availability, 208
Data Integrity, 208
Data Location, 208
Data management, 129
Data Privacy, 208
Data recovery, 230
Data Transmission, 207
De-commissioning, 150
DEA, 113
DeltaCloud, 187
Deployment Models, 4
Detective controls, 223
Deterrent controls, 222
DGC, 24
Distributed Computing, 19
Domain 0, 69
Domain U, 69
Dropbox, 108
Dynamic scaling, 88

EJB container, 18
Elasticity, 130
Emulation, 65
Encapsulation, 67
Eucalyptus, 188
Eucalyptus Systems, 7

Failure-insulated Zone, 88
FCR, 145
FHE, 211
Full virtualization, 73

GAE, 213

Printed in the United States
by Baker & Taylor Publisher Services